Review of NORTHEAST FISHERY Stock Assessments

Committee to Review Northeast Fishery Stock Assessments

Ocean Studies Board

Commission on Geosciences, Environment, and Resources

National Research Council

NATIONAL ACADEMY PRESS
Washington, DC 1998

NOTICE: The project that is the subject of this report was approved by the Governing Board of the National Research Council, whose members are drawn from the councils of the National Academy of Sciences, the National Academy of Engineering, and the Institute of Medicine. The members of the committee responsible for the report were chosen for their special competencies and with regard for appropriate balance.

This report has been reviewed by a group other than the authors according to procedures approved by a Report Review Committee consisting of members of the National Academy of Sciences, the National Academy of Engineering, and the Institute of Medicine.

This report and the committee were supported by the National Oceanic and Atmospheric Administration. The views expressed herein are those of the authors and do not necessarily reflect the views of the sponsor.

This study was supported by Contract No. 50-DKNA-6-90040 between the National Academy of Sciences and the National Oceanic and Atmospheric Administration. Any opinions, findings, conclusions, or recommendations expressed in this publication are those of the author(s) and do not necessarily reflect the view of the organizations or agencies that provided support for the project.

Cover art, by Winslow Homer, titled "The Fog Warning." Provided by the Otis Norcross Fund, courtesy of the Museum of Fine Arts, Boston, Massachusetts.

Library of Congress Catalog Card Number 98-84912
International Standard Book Number 0-309-06030-3

Additional copies of this report are available from:

National Academy Press
2101 Constitution Avenue, N.W.
Box 285
Washington, DC 20055
800-624-6242
202-334-3313 (in the Washington Metropolitan Area)
http://www.nap.edu

Copyright 1998 by the National Academy of Sciences. All rights reserved.

Printed in the United States of America.

COMMITTEE ON NORTHEAST FISHERY STOCK ASSESSMENTS

TERRANCE J. QUINN II, *Chair*, University of Alaska Fairbanks, Juneau
WYATT ANDERSON, University of Georgia, Athens
WAYNE M. GETZ, University of California, Berkeley
RAY HILBORN, University of Washington, Seattle
CYNTHIA JONES, Old Dominion University, Norfolk, Virginia
JEAN-JACQUES MAGUIRE, Advice in Fisheries Science and Management, Quebec, Canada
ANA PARMA, International Pacific Halibut Commission, Seattle
TORE SCHWEDER, University of Oslo, Norway
GUNNAR STEFANSSON, Marine Research Institute, Reykjavik, Iceland

Staff

M. ELIZABETH CLARKE, Study Director
GLENN MERRILL, Research Associate
SHARI MAGUIRE, Project Assistant
ANN CARLISLE, Project Assistant

Consultant

GRAEME PARKES, MRAG Americas

OCEAN STUDIES BOARD

KENNETH BRINK, *Chair*, Woods Hole Oceanographic Institution, Massachusetts
ALICE ALLDREDGE, University of California, Santa Barbara
DAVID BRADLEY, Pennsylvania State University, State College
WILLIAM CURRY, Woods Hole Oceanographic Institution, Massachusetts
ELLEN DRUFFEL, University of California, Irvine
RANA FINE, University of Miami, Florida
CARL FRIEHE, University of California, Irvine
ROBERT GAGOSIAN, Woods Hole Oceanographic Institution, Massachusetts
SUSAN HANNA, Oregon State University, Corvallis
JOHN HOBBIE, Marine Biological Laboratory, Woods Hole, Massachusetts
EILEEN HOFMANN, Old Dominion University, Norfolk, Virginia
JOHN KNAUSS, University of Rhode Island, Narragansett
ROBERT KNOX, University of California, San Diego
RAY KRONE, University of California, Davis
LOUIS LANZEROTTI, Bell Laboratories, Lucent Technologies, Murray Hill, New Jersey
JOHN MAGNUSON, University of Wisconsin, Madison
WILLIAM MERRELL, The H. John Heinz III Center for Science, Economics and the Environment, Washington, D.C.
B. GREGORY MITCHELL, University of California, San Diego
NEIL OPDYKE, University of Florida, Gainesville
MICHAEL ORBACH, Duke University Marine Laboratory, Beaufort, North Carolina
TERRANCE QUINN, University of Alaska Fairbanks, Juneau
C. BARRY RALEIGH, University of Hawaii, Honolulu
JAMES RAY, Shell Oil Company, Houston, Texas
GEORGE SOMERO, Stanford University, Pacific Grove, California
PAUL STOFFA, University of Texas, Austin

Staff

MORGAN GOPNIK, Director
M. ELIZABETH CLARKE, Associate Director
EDWARD R. URBAN, JR., Program Officer
DAN WALKER, Program Officer
ROBIN MORRIS, Financial Associate
GLENN MERRILL, Research Associate
LORA TAYLOR, Senior Project Assistant
JENNIFER SWERDA, Project Assistant
SHARI MAGUIRE, Project Assistant
ANN CARLISLE, Project Assistant

COMMISSION ON GEOSCIENCES, ENVIRONMENT, AND RESOURCES

GEORGE M. HORNBERGER, *Chair*, University of Virginia, Charlottesville
PATRICK R. ATKINS, Aluminum Company of America, Pittsburgh, Pennsylvania
JAMES P. BRUCE, Canadian Climate Program Board, Ottawa, Ontario
WILLIAM L. FISHER, University of Texas, Austin
JERRY F. FRANKLIN, University of Washington, Seattle
THOMAS E. GRAEDEL, Yale University, New Haven, Connecticut
DEBRA KNOPMAN, Progressive Foundation, Washington, D.C.
KAI N. LEE, Williams College, Williamstown, Massachusetts
PERRY L. MCCARTY, Stanford University, California
JUDITH E. MCDOWELL, Woods Hole Oceanographic Institution, Massachusetts
RICHARD A. MESERVE, Covington & Burling, Washington, D.C.
S. GEORGE PHILANDER, Princeton University, New Jersey
RAYMOND A. PRICE, Queen's University at Kingston, Ontario
THOMAS C. SCHELLING, University of Maryland, College Park
ELLEN SILBERGELD, University of Maryland Medical School, Baltimore
VICTORIA J. TSCHINKEL, Landers and Parsons, Tallahassee, Florida
E-AN ZEN, University of Maryland, College Park

Staff

ROBERT HAMILTON, Executive Director
GREGORY SYMMES, Assistant Executive Director
JEANETTE SPOON, Administrative Officer
SANDI FITZPATRICK, Administrative Associate
MARQUITA SMITH, Administrative Assistant/Technology Analyst

The National Academy of Sciences is a private, nonprofit, self-perpetuating society of distinguished scholars engaged in scientific and engineering research, dedicated to the furtherance of science and technology and to their use for the general welfare. Upon the authority of the charter granted to it by the Congress in 1863, the Academy has a mandate that requires it to advise the federal government on scientific and technical matters. Dr. Bruce Alberts is president of the National Academy of Sciences.

The National Academy of Engineering was established in 1964, under the charter of the National Academy of Sciences, as a parallel organization of outstanding engineers. It is autonomous in its administration and in the selection of its members, sharing with the National Academy of Sciences the responsibility of advising the federal government. The National Academy of Engineering also sponsors engineering programs aimed at meeting national needs, encourages education and research, and recognizes the superior achievements of engineers. Dr. William A. Wulf is president of the National Academy of Engineering.

The Institute of Medicine was established in 1970 by the National Academy of Sciences to secure the services of eminent members of appropriate professions in the examination of policy matters pertaining to the health of the public. The Institute acts under the responsibility given to the National Academy of Sciences by its congressional charter to be an adviser to the federal government and, upon its own initiative, to identify issues of medical care, research, and education. Dr. Kenneth I. Shine is president of the Institute of Medicine.

The National Research Council was organized by the National Academy of Sciences in 1916 to associate the broad community of science and technology with the Academy's purposes of furthering knowledge and advising the federal government. Functioning in accordance with general policies determined by the Academy, the Council has become the principal operating agency of both the National Academy of Sciences and the National Academy of Engineering in providing services to the government, the public, and the scientific and engineering communities. The Council is administered jointly by both Academies and the Institute of Medicine. Dr. Bruce M. Alberts and Dr. William A. Wulf are chairman and vice chairman, respectively, of the National Research Council.

CONTENTS

EXECUTIVE SUMMARY   1

1   INTRODUCTION   7
The Groundfish Resources of the Northwestern Atlantic, 7
Historical Exploitation and Assessment, 7
Modern Exploitation and Assessment, 9
Description of 1997 U.S. Stock Assessment for Cod, Haddock, and
    Yellowtail Flounder, 14
History and Purpose of the Study, 23

**2   GENERAL REVIEW OF NORTHEAST GROUNDFISH STOCK
ASSESSMENTS**   25
Investigation of the Northeast Fishery Stock Assessments, 30
Comparison with Assessments Around the World, 34
General Evaluation of Northeast Groundfish Stock Assessment Based on
    Recommendations from *Improving Fish Stock Assessments*, 39
Status of the Five Stocks, 40

**3   REVIEW OF DETAILS OF NORTHEAST GROUNDFISH STOCK
ASSESSMENTS**   61
Data, 61
Assessment Models, 66

4   SCIENCE AND MANAGEMENT   69
Role of Stock Assessment in the Fishery Management Process, 69
Strategic Thinking and Process, 71
Public Hearing, 72

5   CONCLUSIONS   75

REFERENCES   81

APPENDIXES   87

A.   Mandate from Magnuson-Stevens Fishery Conservation Management Act, 89
B.   Committee Biographies, 90
C.   Materials Received from National Marine Fisheries Service, 92
D.   Presentation to the committee by NMFS scientists, 104
E.   Glossary, 116
F.   Extending Data Series and Alternative Projection Results for Gulf of Maine Cod, 122

EXECUTIVE SUMMARY

Groundfish such as cod, haddock, and flounder have been the mainstay of a fishing industry in the Northeast for four centuries. Today many of the groundfish stocks in the northeastern United States and the Canadian Atlantic are estimated to be severely depleted. In the early 1990s, the National Marine Fisheries Service (NMFS) stock assessments suggested that, in particular, five stocks (Gulf of Maine and Georges Bank cod, Georges Bank haddock, and Georges Bank and southern New England yellowtail flounder) shared similar characteristics: low spawning stock biomass relative to 20 years before and high fishing mortality rates, with 50-80 percent of the fish being captured every year. Canadian stocks were similarly assessed at record lows by the Canadian Department of Fisheries and Oceans (DFO).

Based on these assessments, NMFS concluded that maintaining such high fishing mortality rates would lead to continued low catches and the likelihood of major collapse of the stocks and catch. In 1995 and 1996, the New England Fishery Management Council and NMFS implemented strong management measures that closed vast coastal areas of the Gulf of Maine and Georges Bank fishing grounds and placed various restrictions on fishing. These restrictions have caused concern among many New England coastal communities that depend on these fisheries for their economic livelihood. In particular, harvesters have questioned whether the scientific information used to make these decisions warranted such strong management actions.

In response to concerns about the stock assessment results, Congress mandated that the National Academy of Sciences conduct a review of Canadian and U.S. stock assessments, information collection methodologies, biological assumptions and projections, and other relevant scientific information used as the basis for conservation and management in the Northeast multispecies fishery (16 U.S.C. 1801 et seq., see Appendix A). The Academy's National Research Council formed the Committee on Northeast Fishery Stock Assessments to address this issue.

Stock assessment is the science of data collection, analysis, and modeling that provides the basis for prudent, sustainable use of fishery resources. Stock assessments include scientific advice about management strategies used to exploit stocks and the integration of science and scientific advice into the management process. Stock assessments, understood narrowly as the collection and interpretation of data necessary to estimate stock status and predict effects on the stocks of a given fishery pattern, are not the only inputs to fishery management. Issues of law and access rights, institutional and regulatory frameworks, policy formation and implementation, ecology and environment are all central to sound fishery management. Because the committee was asked to focus this review on specific scientific issues described in its congressional mandate, such broad issues will not be addressed explicitly in this report. **The committee does recognize, however, that it is not enough to have an excellent stock assessment program: achieving a sustainable fishery will require an equally strong management system that takes into account the issues identified above.**

The committee concentrated its review on the stock assessments for cod, haddock, and yellowtail flounder. In 1997, the Canadian and U.S. stock assessments for these stocks were conducted collaboratively by the Canadian Department of Fisheries and Oceans and NMFS. The committee's approach was to (1) attempt to replicate NMFS stock assessments; (2) compare the current stock assessment with other assessments around the world; (3) evaluate the stock assessment of cod, haddock, and yellowtail flounder according to the guidelines outlined in the report *Improving Fish Stock Assessments* (NRC, 1998); (4) evaluate the status of the stocks of cod, haddock, and yellowtail flounder; and (5) evaluate the role of stock assessment science in the fishery management process.

ADEQUACY OF SCIENTIFIC INFORMATION FOR MANAGEMENT

Improving Fish Stock Assessments (NRC, 1998) reviewed the state of existing knowledge about the stock assessment process. The report stressed that the feedback between stock assessment and fisheries management has to be improved to manage fisheries more effectively. The report also includes nine recommendations pertinent to stock assessment. The committee examined these recommendations in relation to the Northeast fishery stock assessments, and concluded that most of the earlier report's recommendations are already being addressed by NMFS in these fisheries. **The current stock assessment process, despite the need for improvements, appears to provide a valid scientific context for evaluating the status of fish populations and the effects of fishery management. Furthermore, the process is analogous to processes used in jurisdictions elsewhere in the world. Therefore, the Northeast stock assessment process is well within the standards of the stock assessments conducted elsewhere in the United States and by other nations.**

In all five stocks considered, fishing mortality was high, increasing and not sustainable, whereas spawning stock biomass was low and decreasing. The available data and the assessments using them show convincingly that the reviewed stocks have been subject to increased fishing mortality and decreasing spawning stock biomass through the 1980s and early 1990s. These conditions considerably increase the risk of major stock collapse. The increasing fishing mortality rates during the 1980s and early 1990s indicate that management measures implemented during that period were ineffective in controlling fishing mortality. Therefore a different, more drastic approach was needed to decrease the probability of stock collapse.

JUSTIFICATION FOR MANAGEMENT ACTIONS

Overall, the demonstrated tenuous status of the five stocks and the substantial uncertainty surrounding their ability to recover warranted strong management action. The committee finds no scientific basis to support assertions that the regulations imposed by Amendment 7 are too severe from a biological perspective. In fact, further management action may be necessary for the Gulf of Maine cod fishery. It should be added that the regulations in Amendments 5-7 might have been avoided if fishing mortality in the New England groundfisheries had been effectively controlled from the mid-1980s.

If stock sizes are at intermediate levels relative to historical values, an argument might be made simply to maximize the yield from recruitment obtained. However, current scientific wisdom indicates that a precautionary approach should be used for harvesting unless there is conclusive evidence that higher harvests will result in better ecosystem regulation. **When stocks are low, as in the Northeast, high fishing mortality markedly increases the risk of irrevocable stock damage.**

In two of the five cases in the Northeast (Georges Bank haddock and southern New England yellowtail flounder), the committee believes that the stocks already have collapsed*, as indicated by low spawning levels combined with a period of little or no recruitment. Recruitment remains low and recent increases in biomass primarily result from higher survival of small year classes and growth of the current biomass because of the lower fishing mortality. For Georges Bank cod, yellowtail flounder, and Gulf of Maine cod, some of the committee's simulations of future stock size show that there would be a real danger of future stock collapse if strong regulations to reduce fishing mortality were not in place. For Gulf of Maine cod, the stock does not appear to have collapsed but there is danger it could under recent fishing mortality. Current regulations have not yet shown the ability to control fishing mortality for this stock. Additional management measures may be required.

EFFECTS OF MANAGEMENT REGULATIONS

The assessments indicate that fishing mortality played a major role in reducing the abundance of groundfish in New England. Current stock assessments suggest that fishing mortality has been reduced for four of the five reviewed stocks and that these stocks appear to be increasing. The fifth stock, Gulf of Maine cod, has not experienced reduced fishing mortality, and it is not increasing. Biomass remains considerably smaller than observed in the past, and any relaxation of management measures may jeopardize sustained stock rebuilding.

SUFFICIENCY OF SURVEY AND LANDINGS DATA

The scope and protocols of current data inputs (trawl survey and landing data) are sufficient for the stock assessments. Improvements in data collection systems are necessary, especially for catch and discard reporting, survey coverage, and collection of age information. Nevertheless, the assessment inputs are comparable to and, to some extent, better than those available for other stocks in the United States and elsewhere. The uncertainty of the stock assessments could be reduced further if the data inputs were improved.

USEFULNESS OF STOCK ASSESSMENTS AS PREDICTIVE TOOLS

There is no simple answer regarding the adequacy and reliability of these stock assessments as predictive tools. The current assessments evaluate how implementing different fishing mortalities (F), including no change in F, would affect the annual probabilities of reaching the rebuilding thresholds set by the New England Fishery Management Council. Future biomass trajectories can only be predicted in probabilistic terms, and there is great uncertainty about how long the stocks will take to rebuild to these thresholds.

While NMFS forecasts do incorporate substantial uncertainty, some of the main sources of uncertainty have been left out. Knowledge of the biology of the fish stocks is incomplete, and this

---

*The committee made a determination of whether a stock had collapsed by examining the historical estimates of spawning biomass and recruitment. See Chapter 2, page 41.

contributes to uncertainty in stock projections. For example, it is unclear whether changes in recruitment of young fish are environmentally controlled or the result of low spawning stock sizes. **Different hypotheses about the relationship between stock abundance and subsequent recruitment are consistent with the data at hand, and they tend to push the time of recovery further out into the future. Under some of these hypotheses, the probability of the stock not having recovered to the Council thresholds, say within 10 years, is thus substantially larger than suggested by NMFS assessments. To estimate such probabilities quantitatively is not easy. It is therefore difficult to base the management decisions directly on predictions from stock assessments, without embodying the assessment and its predictions in a management strategy with precautionary qualities.**

ALTERNATIVE METHODS TO REGULATE CATCH

Technical measures, such as mesh size increases, were used extensively in the Northeast groundfish fisheries during the 1980s and early 1990s, and they did not succeed in limiting fishing mortality. Mesh size restrictions often do not lead to conservation of fish stocks unless fishing mortality itself is somehow controlled, for example, through setting a Total Allowable Catch (TAC) for the species, establishing closed areas, or limiting fishing effort. Increases in the closed areas appear to have played an important role in reducing fishing mortality in 1995 and 1996. The New England Fishery Management Council may choose to adjust mesh size and closed areas to fine-tune the control of fishing mortality. One problem with so-called technical measures is that they make it harder to quantitatively determine the effects of management actions on fishing mortality and stock size. **A valuable role for future stock assessments will be to investigate whether and how the effects of these management actions on both fishing mortality and stock size can be measured.**

FACTORS AFFECTING ABUNDANCE OF STOCKS

Evaluating the relative contributions of different factors in driving changes in the abundance of fish stocks is difficult, but as mentioned above, fishing mortality played a major role in reducing the abundance of the five stocks in the 1980s and early 1990s, and still does for the Gulf of Maine cod stock. The most recent estimates of fish mortality are the most uncertain, and assessments in future years may come up with much different estimates for 1995 and 1996. It should be noted that hydroclimatic changes have occurred in U.S. waters of the northwestern Atlantic, but their magnitude is substantially less than off Newfoundland and Labrador, where they are believed to be one of the causative factors in stock collapses. **However, even if environmental changes were a factor affecting stock abundance, fishing mortalities were high enough to contribute to major stock collapses. To avoid stock collapse it is more important to stringently reduce fishing mortality during periods of adverse environmental conditions.**

Article 6.2 of the UN Agreement on Highly Migratory Fish Stocks and Straddling Fish Stocks (FAO, 1995) directs that: "States shall be more cautious when information is uncertain, unreliable or inadequate. The absence of adequate scientific information shall not be used as a reason for postponing or failing to take conservation and management measures." Further, Article 6.7 of this agreement specifies that: "If a natural phenomenon has a significant adverse impact on the status of [straddling fish stocks or highly migratory] fish stocks, States shall adopt conservation and management measures on an emergency basis to ensure that fishing activity does not exacerbate such adverse impact. States shall also adopt such measures on an emergency basis where fishing activity presents a serious threat to the sustainability of such stocks."

OTHER CAUSES OF STOCK FLUCTUATIONS

There is speculation among some interested parties that pollution and habitat destruction are reducing stock recruitment. If, in fact, these factors have affected recruitment (and there is no consensus in the scientific community), then this would not alter the scientific recommendation to reduce harvest levels. The fishing mortality recommended for stock rebuilding would have to be adjusted downward for decreased recruitment. Furthermore, if these factors are likely to continue for a long time, then it may not be possible to rebuild stocks to levels which occurred in the past. The larger issue of establishing a causal mechanism between stock fluctuations and environmental and pollution factors is a difficult one because all factors are changing at the same time and so their effects are confounded. However, future directed research in this area, involving environmental and ecosystem studies and cross-population comparisons of key stock-recruitment parameters, should be very valuable for constructing likely scenarios for policy evaluation.

RECOMMENDED ACTIONS

The committee recommends that the National Marine Fisheries Service take the following actions to improve the Northeast stock assessments:

1. Improve the collection, analysis, and modeling of stock assessment data as detailed in Chapter 3. Such improvements could include evaluations of sample size, design, and data collection in the fishery and the surveys; the use of alternative methods for data analysis; consideration of a wider variety of assessment models; and better treatment of uncertainty in forecasting;
2. Improve relationships and collaborations between NMFS and harvesters by providing, for example, an opportunity to involve harvesters in the stock assessment process and using harvesters to collect and assess disaggregated catch per unit effort data;
3. Continue to educate stock assessment scientists through short-term exchanges among NMFS centers so that each center can keep abreast of the latest improvements in stock assessment technologies being used at other NMFS fishery science centers and other organizations in the United States or elsewhere;
4. Ensure that a greater number of independent scientists from academia and elsewhere participate in the Stock Assessment Review Committee (SARC) process; where necessary, pay competitive rates for such outside participation to ensure that a sufficient number of the best people are involved in the review;
5. Increase the frequency of stock assessments. As the New England Fishery Management Council intensifies its management of the Northeast fishery, stock assessments may have to be performed more frequently than every three years (the current timing);
6. Consider a wider range of scenarios (e.g., recruitment, individual growth, survival, sub-stock structure, ecosystem, data quality, compliance with regulations, long-term industry response) in evaluating management strategies;
7. Investigate the effects of specific management actions, such as closed areas and days at

sea limitations, on fishing mortalities and related parameters;

8. Work toward a comprehensive management model that links stock assessments with ecological, social and economic responses, and adaptation for given long-term management strategies. This involves input from the social sciences (economics, social and political science, operations research) and from a wider range of natural sciences (ecology, genetics, oceanography) than traditionally is the case in fisheries management.

The committee has not explicitly considered the costs of implementing these recommendations, which may require either additional resources or a reprogramming of existing resources.

The committee concludes that stock assessment science is not the real source of contention in the management of New England groundfish fisheries. Comments at a public hearing held by the committee support this conclusion. Many speakers suggested that the social and economic concerns created by strong management measures and lack of participation in the management process were the more important concerns. Traditional fishery science has a major role to play in fisheries management, but sound stock assessment clearly is not the only consideration.

The New England Fishery Management Council will be facing critical decisions, depending on the recovery or non-recovery of groundfish stocks. A long-term management strategy will be needed to decide the rate of rebuilding required to reach particular targets. Without sound stock assessment, targets and rebuilding rates cannot be set, nor can the effectiveness of the regulatory actions be measured. However, stock assessment in the narrow sense of estimating status and dynamics of fish populations is not sufficient for rational fisheries management.

What constitutes a good management approach will vary over time, location, and components of the fish stock. To obtain the information necessary to design effective institutional and regulatory frameworks, it is essential that management draws on stock assessment, oceanography, ecology, economics, social and political science and operations research. **Only when a more comprehensive approach is taken, with long-term management strategies based on data and insight from the various fields, properly accounting for the uncertainties surrounding data and theory, can fishery management provide for high continuing yield of food and health of stocks, while considering the needs of people dependent upon the fisheries.**

1

INTRODUCTION

If we only knew all the laws of Nature, we should need only one fact, or the description of one actual phenomenon, to infer all the particular results at that point. Now we know only a few laws, and our result is vitiated, not, of course, by any confusion or irregularity in Nature, but by our ignorance of essential elements in the calculation.

Henry David Thoreau

GROUNDFISH RESOURCES OF THE NORTHWESTERN ATLANTIC

Fisheries for groundfish off the shores of the Northeast United States and Southeast Canada have played a major role in the development of regional commerce, trade, and society in the United States and Canada. Cod fishing became a major source of income and food for the early colonists (Kurlansky, 1997). The export of salt cod to Europe, and later to the West Indies, in the infamous "golden triangle" trade of cod, rum, and slaves, provided hard currency for the developing nation (Jensen, 1972). In the past 10 years, many traditionally important stocks have fallen to their lowest levels in recorded history and, consequently, many of the fishing communities that relied on these once rich fisheries for their economic livelihood have experienced reduced landings.

Attempts to limit the current fishing effort and implement other restrictive regulations in order to promote the rebuilding of these stocks have led to major socioeconomic, ecological, and political concerns for coastal communities in New England. The well-being of fish and harvesters is likely to continue to be a critical issue for fishery management in the Northeast in the future. In particular, three of the most important commercial fish species, cod (*Gadus morhua*), haddock (*Melanogrammus aeglefinus*), and yellowtail flounder (*Pleuronectes ferruginea*), have all experienced substantial reductions in catch and stock (Figures 1.1 and 1.2).

HISTORICAL EXPLOITATION AND ASSESSMENT

Early groundfish fisheries, characterized by small fishing vessels that targeted cod, were largely unregulated until this century. These fisheries were located in inshore waters, but trips offshore to Georges Bank began in the mid-1700s. Georges Bank became a principal harvesting ground by the late 1800s (Jensen, 1972). Prior to the early 1900s, most cod were captured by handlining from schooners or longlining from small dories that delivered their catch to larger schooners at the end of the day. By using these simple, but labor-intensive methods, nearly 50,000 tons of cod were harvested from Georges Bank in 1893. This was the first year for which a relatively accurate record of commercial landings was kept (Serchuk and Wigley, 1992).

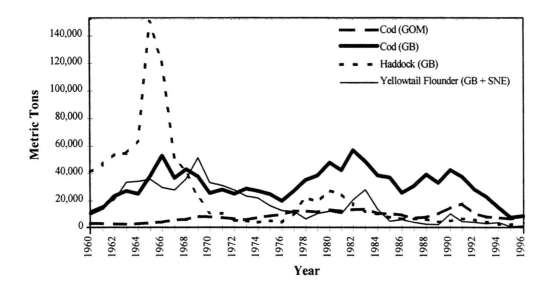

FIGURE 1.1 Commercial Landings (metric tons, live) of cod, haddock, and yellowtail flounder from Gulf of Maine (GOM) (NAFO Division 5Y), Georges Bank (GB) and southern New England (SNE) (NAFO Division 5Z and Subarea 6). SOURCE: NEFSC, 1997a. NOTE: NAFO = Northwest Atlantic Fisheries Organization.

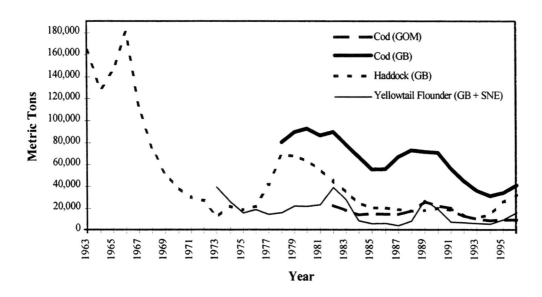

FIGURE 1.2 Spawning stock biomass (metric tons) at the start of the spawning season of cod, haddock, and yellowtail flounder from Gulf of Maine (GOM), Georges Bank (GB), southern New England (SNE). Based on ADAPT-tuned VPA. SOURCE: NEFSC, 1997a. NOTE: VPA = virtual population analysis.

Introduction

The groundfish fishery developed rapidly with the introduction of steam- and diesel-powered vessels, otter trawling, and on-board refrigeration around the turn of the century (Clark et al., 1982; Serchuk and Wigley, 1992; Kurlansky, 1997). Otter trawling, which used large planks or doors to pull the net open, was developed in England and introduced to the Northeast fishery in 1905 (Jensen, 1972). This technique, when combined with motorized vessels, was much more efficient than traditional dory fishing, and fishing effort and fleet size in the Northeast groundfish fishery increased rapidly after its introduction (Jensen, 1972; Clark et al., 1982). However, fishing efforts in the Northeast declined, and heavily targeted cod stocks appeared to recover due to the reduction in fishing during World War I. Between World Wars I and II, effort in the groundfish fishery again increased and shifted from cod to Georges Bank haddock. The harvest of haddock peaked at 132,200 tons in 1929, then dropped to 28,000 tons by 1934 (Clark et al., 1982).

In the mid-1930s, another commercially important fishery developed around yellowtail flounder. The winter flounder that had previously supported the New England trawl fleet declined drastically, and yellowtail flounder became a substitute for trawlers and filleting plants (Royce et al., 1959). The higher demand for food during World War II increased fishing effort generally and, specifically, effort on yellowtail flounder. Yellowtail flounder quickly became an important target species, and harvesting rates rose rapidly: a peak of nearly 29,000 metric tons was harvested in the southern New England fishery in 1942 (Royce et al., 1959). The effort in this fishery increased and the price per pound received by harvesters in 1947 had more than doubled since 1942 (Royce et al., 1959). The stock subsequently collapsed, and less than 15,000 tons were harvested in the southern New England fishery in 1949.

An international convention established the International Commission for the Northwest Atlantic Fisheries (ICNAF) in 1949 to provide a forum for the management of northwest Atlantic fish stocks shared by more than one nation. A method of dividing fishing grounds into statistical areas based on ecological and oceanographic characteristics was formalized under ICNAF, and critical data were organized by managers (Hennemuth and Rockwell, 1987).

The exploitation of cod, haddock, and yellowtail flounder by distant-water factory trawler fleets from the Soviet Union, Spain, and other European countries began in 1961 and expanded rapidly in the early and late 1960s (Clark et al., 1982; Serchuk and Wigley, 1992; Murawski et al., 1997). Fishing effort increased and cod landings rose more than fivefold, from 11,000 tons in 1960 to 53,000 tons in 1966 (Serchuk and Wigley, 1992). The catch of northern cod stocks off Newfoundland and Labrador peaked at 810,000 metric tons in 1968 then diminished rapidly under similarly intensive fishing by distant-water fleets (Hutchings and Myers, 1994).

Standardized catch sampling was initiated for haddock in 1931, and for other species on a case-by-case basis thereafter, until 1963 when a multispecies sampling program was implemented (Serchuk and Wigley, 1992). In the same year, surveys were conducted by the United States in autumn, and by 1968, they had been expanded to include both autumn and spring surveys.

MODERN EXPLOITATION AND ASSESSMENT

The arena for fisheries management changed drastically in the 1970s. Depletion of fish stocks by the foreign fishing fleets off U.S. coasts, in part, motivated the United States to extend its fishing jurisdiction out to 200 nautical miles from shore. The need to manage fisheries in this extended jurisdiction and to improve the management of rapidly declining fish stocks led to new federal legislation. The Fishery Conservation and Management Act of 1976 (P.L. 94-265, 16 U.S.C. 1801 et seq.) established regions within which fish populations would be managed and mechanisms for controlling fishing activities. The Magnuson Fishery Conservation and Management Act (MFCMA), as it was later renamed, called for the formation of regional fishery management councils. The New England Fishery Management Council (NEFMC) was formed in 1977 and the modern era of management of the groundfish stock off the northeastern coast began. Recent stock assessment activities and actions are summarized in Table 1.1.

Although joint U.S. and Canadian management of shared Georges Bank and Gulf of Maine groundfish stocks had been somewhat successful under ICNAF, cooperative management faltered quickly due to disagreements over national boundaries. By 1978, each country set total allowable catches (TACs) individually, although scientific information continued to be shared among fishery scientists.

The NEFMC approved the first fishery management plan (FMP) for the Northeast groundfish fishery in 1977 (Serchuk and Wigley, 1992). However, it quickly became apparent that groundfish stocks were not rebuilding (Serchuk and Wigley, 1992). The individual trip limit system that had been implemented to reduce fishing effort had proven unsuccessful due to difficulties in accurately monitoring daily landings of vessels, misreporting of areas in which fish were caught, illegal landings at many ports along the coast, and mislabeling of fish species (NEFMC, 1994). NEFMC began work on a new interim plan to replace the existing FMP in 1978 (Serchuk and Wigley, 1992). In 1982, NEFMC abandoned the use of the TAC quota system established by ICNAF, in favor of indirect effort control measures such as minimum mesh size restrictions, area closures, and minimum fish size regulations for cod, haddock, and yellowtail flounder (Serchuk and Wigley, 1992; Clark et al., 1982; NEFSC, 1994b).

Canadian efforts to improve stock assessment methods increased with its extended jurisdiction to 200 miles offshore. In 1977, the Canadian Atlantic Fisheries Scientific Advisory Committee (CAFSAC) was formed to provide scientific advice for the management of Canadian groundfish stocks. The exclusive economic zones (EEZs) of the United States and Canada overlapped in the region of Georges Bank. Unfortunately, the allocation of fishing privileges within this zone could not be negotiated between the two nations, and the United States and Canada submitted to binding arbitration by the International Court of Justice (ICJ). In 1984, a maritime boundary was determined by the ICJ, and implemented by the two nations in 1985.

The first Canadian assessment of Georges Bank cod was conducted in 1983, and annual assessments have been performed in most years since (Hunt and Buzeta, 1996). Canadian harvests increased until the mid-1980s, after which they declined rapidly (Hutchings and Myers, 1994). In July 1992, the northern cod fishery off Newfoundland and Labrador was closed. Canadian management had been based primarily on setting a single-species annual TAC, but it also includes other methods of control such as seasonal and area closures (Mayo et al., 1992). Since 1992, Canada has dramatically altered the management of Atlantic groundfish (Fordham, 1996). The most significant changes have been the implementation of a strict dockside monitoring program, an enhanced at-sea observer program, and an individual transferable quota (ITQ) program for many stocks.

In 1988, environmental groups and the Technical Monitoring Group established by NEFMC to review the multispecies groundfish FMP raised concerns about the status of groundfish stocks and the effectiveness of management in rebuilding stocks (Serchuk and Wigley, 1992). In 1991, the Conservation Law Foundation and the Massachusetts Audubon Society sued the National Marine Fisheries Service (NMFS) for failing to prevent overfishing as defined in Amendment 4 to the multispecies groundfish FMP implemented in January 1991. A consent decree was reached between the parties in 1991, requiring that the overfishing levels for groundfish species established under Amendment 4 be met, and the implementation of a new plan to rebuild cod and yellowtail flounder stocks within five years and the haddock stock within ten years (NEFMC, 1994). NEFMC began to develop a new amendment to the groundfish multispecies FMP soon after Amendment 4 was implemented to increase the declining biomass of important commercial stocks, and to meet the requirements of the consent decree.

TABLE 1.1 Management Actions and Stock Assessments for Cod, Haddock, and Yellowtail Flounder from 1960 to 1997

Year	Cod	Haddock	Yellowtail Flounder
1960	—First modern recreational survey in Northeast—		
1963	—Standardized research vessel multispecies trawl surveys (autumn)—		
Early 1960s	—Commercial fishery weigh-out, interview, and catch sampling (ICNAF)—		
1965	—Second recreational survey—		
1965-1969	—Peak harvests by foreign distant-water fleets—		
1968	—Standardized research vessel multispecies trawl surveys (spring and autumn)—		
1970		Spawning area closures from March-April. ICNAF sets TAC at 12,000 tons (Georges Bank)	
1971	Assessment indicates Subarea 5 cod exceeded MSY since 1965	—First formal stock assessment under ICNAF—	
1972		Spawning area closures March-May. ICNAF sets TAC at 6,000 tons (Georges Bank)	
1973	ICNAF sets TAC of 35,000 tons		
1975			Spawning area closures February-May. ICNAF sets incidental catch only TAC
1976	VPA-based recommendation of 15,000-ton TAC	—Magnuson Fishery Management and Conservation Act (MFCMA) passed— —Preliminary VPA conducted—	

TABLE 1.1 (continued)

Year	Cod	Haddock	Yellowtail Flounder
1977		—Fishery Management Council established—	
		—Reciprocal Fishing Agreement (U.S.-Canada)—	
		—CAFSAC formed and $F_{0.1}$ established as long-term harvesting goal (Canada)—	
		—TAC and indirect effort control measures used (Canada)—	
1978	VPA conducted using commercial and recreational data. Indicates high F (0.55-0.65) and low biomass		
		—Groundfish Fishery Management Plan (FMP)—	
		—Minimum mesh size 5.125"—	
		—Trip limits, mandatory reporting, quotas set, recreational catch under FMP—	
		—End of Reciprocal Fishing Agreement—	
	Minimum fish size 16"		
1979		—Recreational surveys standardized and conducted annually—	
		—"Interim Plan" for groundfish—	
		—Catch controls eliminated, closed areas, minimum fish and mesh sizes retained—	
1982	Minimum fish size 17" (15" recreational) Peak Canadian harvest of 18,000 tons under TAC		
1983		—First Canadian assessment of Georges Bank multispecies complex—	
		—SPA performed using 1960-1976 data (Canada)—	
		Minimum fish size 17" (15" recreational) Minimum mesh size 5.5"	Minimum fish size 11"
1984		—International Court of Justice ends U.S.-Canada boundary dispute—	
		—Court delineates Hague Line—	
1985		—Multispecies FMP implemented—	
		—Increased gear and effort controls—	
	Minimum mesh size 5.5"		Minimum mesh size 5.5"
1986		Area closures extend from February-May	Area closures enacted in southern New England
1987		—Canadian SPA updated using ADAPT model—	

TABLE 1.1 (continued)

Year	Cod	Haddock	Yellowtail Flounder
1988	—Canadian DFO suggests improved U.S.-Canada controls—		
	—VPA updated—		
			Minimum fish size 12"
1989	—Canada establishes new management efforts, focuses efforts on cod—		
	—VPA updated—		
1991	—Amendment 4—		
	—CLF and Massachusetts Audubon sue NMFS for lack of enforcement of overfishing definition—		
	—Consent decree (NMFS) to prevent overfishing and rebuild stocks—		
	—5-year (cod and yellowtail) and 10-year (haddock) rebuilding stocks—		
	—Stricter effort control measures enacted (area closures increased)—		
	—Canada introduces ITQs and dockside monitoring of catch—		
	—Area closures extended—		
1992	—U.S. and Canadian stock assessments completed—		
	—Canada increases at-sea monitoring—		
1994	—Canada establishes minimum catch history for ITQ vessels in Georges Bank—		
	—Amendment 5—		
	—Amendment 6—		
	—Area closures in key areas now year round—		
1996	—Amendment 7—		
1997	—U.S. and Canadian stock assessments completed—		

NOTE:

CAFSAC = Canadian Atlantic Fisheries Scientific Advisory Committee
CLF = Conservation Law Foundation
DFO = Department of Fisheries and Oceans
FMP = fishery management plan
ICNAF = International Commission for Northwest Atlantic Fisheries

ITQ = individual transferable quota
MSY = maximum sustainable yield
NMFS = National Marine Fisheries Service
SPA = sequential population analysis
TAC = total allowable catch
VPA = virtual population analysis

In 1993, NMFS reported that the haddock stock on Georges Bank was the lowest on record and that landings of Gulf of Maine haddock had decreased by 96% from 1983 to1992 (Fordham, 1996). Emergency regulations were implemented. These included haddock closed area regulations, prohibition of transfer of fish at sea, and a ban on pair trawling. Some of these emergency measures were included the following year in Amendment 5. Those that were not included in this amendment became part of Amendment 6, prepared by the Secretary of Commerce.

In 1994, NEFMC produced Amendment 5 (Table 1.2). This new amendment called for a 50 % reduction in fishing effort divided equally over a five- to seven-year period (MMC, 1996). This effort reduction was to be accomplished by a moratorium on new permits for the groundfish fishery, stricter area closures (Figure 1.3), minimum mesh and fish sizes, a mandatory logbook reporting system, and a days at sea (DAS) program (MMC, 1996). The DAS program was designed to reduce effort by limiting the number of days that a fishing vessel could fish. The number of days allocated to a vessel was determined by one of two methods depending on the preference of the vessel owner (1) as a fraction of that vessel's historical effort or (2) by assigning a "fleet" value, based on averaged values from the groundfish fleet (MMC, 1996). Amendment 5 specified a five-year schedule for reducing DAS by 10 percent per year for vessels under the individual vessel DAS program with the goal of a 50% reduction in by 1999 (Table 1.2). This reduction translates into 88 DAS in 1999 (MMC, 1996).

In 1994, the United States conducted a stock assessment for Northeast groundfish stocks. This assessment concluded that Georges Bank and Gulf of Maine cod, haddock, and yellowtail flounder stocks were at or near record low levels of spawning biomass and that cod and yellowtail flounder fishing mortality levels were far in excess of fishing mortality levels needed to rebuild the stocks (NEFSC, 1994a, 1995; O'Brien and Brown, 1995). Canadian assessments from 1995 indicated similarly low levels of spawning biomass for stocks on Georges Bank (Gavaris and VanEeckhaute, 1996; Hunt and Buzeta, 1996).

The Stock Assessment Review Committee (SARC) convened by NMFS in June 1994 concluded that "measures provided in Amendment 5 are clearly inadequate" (NEFSC, 1994b) and that fishing mortality for cod and yellowtail "should be reduced to levels approaching zero" (NEFSC, 1994b). Amendment 7 was implemented in July 1996 and is similar in many respects to Amendment 5. However, Amendment 7 accelerated the reductions in total fleet DAS to 88 days in 1997 and reduced individual vessel DAS by 50% by 1997 (MMC, 1996). For the first time since 1982, Amendment 7 set a target TAC for groundfish species (MMC, 1996). This target provided a recommended catch, although reaching the target TAC would not necessarily trigger management action as TACs in other fisheries do. Amendment 7 increased the number and duration of area closures (Figure 1.4), reduced the number of exemptions to the DAS program, and established a Multispecies Monitoring Committee (MMC). The MMC was to review progress toward meeting the fishing mortality reduction goals established under provisions of Amendment 7.

DESCRIPTION OF 1997 U.S. STOCK ASSESSMENT FOR COD, HADDOCK, AND YELLOWTAIL FLOUNDER

Terms of Reference

By the end of 1996, the status of Gulf of Maine cod, Georges Bank cod, haddock, and yellowtail flounder, and southern New England yellowtail flounder fisheries was of major concern. Subsequently, the scheduled stock assessments took on a special significance. The terms of reference for these stock assessments were (1) assess the stock status through 1996 and characterize the variability of estimates of stock abundance and fishing mortality rates; (2) provide projected estimates of catch for 1997-1998

Introduction

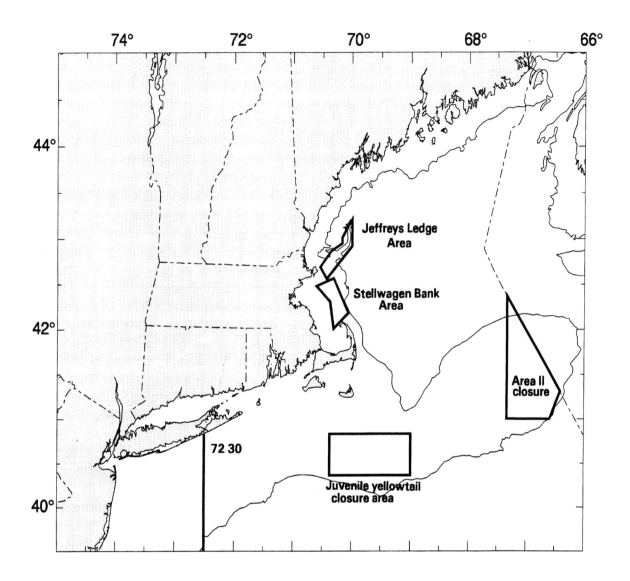

FIGURE 1.3 Regulated areas in the northeast fishery recommended in Amendment 5 of the Multispecies Fisheries Management Plan. SOURCE: NEFMC, 1994.

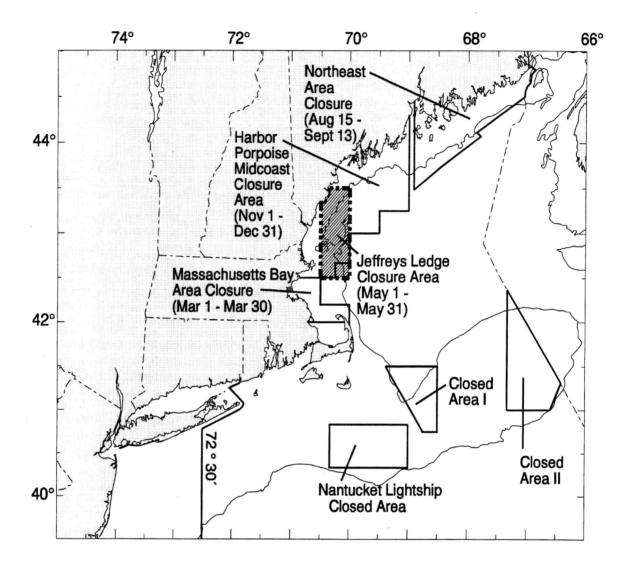

FIGURE 1.4 Regulated areas in the northeast fishery recommended in Amendment 7 of the Multispecies Fisheries Management Plan. SOURCE: NEFMC, 1996.

Introduction

TABLE 1.2 Management Measures Implemented by Amendments 5, 6, and 7.

Amendment 5 (1994)	Amendment 6 (1994)	Amendment 7 (1996)
Moratorium on new vessel permits during stock rebuilding period. Vessels retaining less than 500 pounds of groundfish, with fewer than 4,500 hooks or fishing with sink gillnet gear, are exempted from the moratorium.		Moratorium on new vessel permits during stock rebuilding period. No exemptions for commercial vessels to the moratorium.
Days at sea (DAS) program enacted to reduce fishing vessel effort to 50% of 1993 levels, 10 % per year, within 5 years (1999) of enactment of the plan by reducing the number of days available for fishing. Vessels less than 45 feet, longline vessels with fewer than 4,500 hooks set per day, and certain types of gillnet vessels exempted.		DAS program accelerated to reduce fishing vessel effort to 50% of 1993 levels within 2 years (1997). Vessels of more than 30 feet, longline vessels with fewer than 4,500 hooks set per day, and certain types of gillnet vessels are no longer exempt from days at sea program. Vessels under 30 feet may fish under small vessel exemption permit.
6-inch minimum mesh size applies to most vessels retaining more than 500 pounds of groundfish, certain exemptions for purse seiners and midwater-trawl vessels. Seasonal 6-inch square mesh for juvenile cod protection on Stellwagen Bank and Jeffreys Ledge from March-July.		6-inch mesh size applies to all vessels, unless the bycatch of regulated groundfish species is less than 5% of the weight of the total catch. No retention of groundfish species allowed. 6-inch minimum mesh size for juvenile cod protection on Stellwagen Bank and Jeffreys Ledge year round. 6-inch minimum mesh size also expanded in southern New England east of 72°30' consistent with other areas.
Mandatory logbook reporting required.		Mandatory logbook reporting required. Target total allowable catch (TAC) established. Multispecies Monitoring Committee (MMC) formed to monitor effectiveness of Amendment 7.
500-pound trip limit for haddock.	500-pound trip limit for haddock.	1,000-pound trip limit for haddock.
Closed areas for haddock, cod, and southern New England yellowtail flounder expanded. Time frames of closures for haddock and yellowtail flounder (Areas I, II, and Nantucket Lightship) expanded from February-May to January-June, except haddock in Closed Area II has February-May closure.	Time frames of closures for haddock and yellowtail flounder (Areas I, II, and Nantucket Lightship) expanded year-round by emergency action in December. Scallop dredging prohibited year-round in closed areas. Scallop dredges prohibited from possessing or landing haddock from January-June. Pair trawling banned.	Closed areas for haddock, cod, and southern New England yellowtail flounder expanded. Time frames of closures for haddock and yellowtail flounder (Areas I, II, and Nantucket Lightship) year-round. Scallop dredging prohibited year-round in closed areas.

and spawning stock biomass (SSB) for 1998-1999 at various levels of F (fishing mortality), including all relevant biological reference points; and (3) advise on the assessment and management implications of incorporating recreational catch and commercial discard data in the assessment.

United States and Canadian Stock Assessment Process

Stock assessments are conducted jointly by teams of scientists from the Department of Fisheries and Oceans of Canada (DFO) and the U.S. National Marine Fisheries Service (NMFS). The assessments are subsequently peer reviewed by federal, state, provincial, and academic scientists. In Canada the assessments are reviewed during the regional advisory process (RAP), and in the United States the stock assessments are reviewed during the SARC process. Final assessment advice is formulated at the last step in the process, the Stock Assessment Workshop (SAW) in the United States, and in Canada, by the Fisheries Resource Conservation Council (FRCC).

Stock Assessment Procedure and Timing in 1997

Scientists at the NMFS Northeast Fisheries Science Center (NEFSC) in Woods Hole, Massachusetts conduct the stock assessments, which incorporate landings, survey, and observer data. The stock assessment process in 1997 started with assembly of data as early as February 1996, preliminary runs of the ADAPT model (Conser and Powers, 1990; Conser et al., 1991) in March 1997, and ended with the final distribution of stock assessment documents to the SARC at its meeting in May 1997 (Figure 1.5). Scientists from NEFSC also participated in the Canadian's regional assessment process in Moncton, New Brunswick, Canada.

The SARC met during the 24th Northeast Regional Stock Assessment Workshop (Figure 1.5). This committee peer-reviewed the stock assessment. The SARC considered the reports of the Northern and Southern Demersal Working Groups (NEFSC, 1997a), peer-reviewed the assessments, and developed management advice. The SARC's advice is published in a report (NEFSC, 1997b) and was presented at meetings of the regional fishery management councils.

Dealer and Vessel Data

The data used in each of the assessments are summarized in Appendix D. Beginning in 1994, personal interviews were no longer used to determine fishing effort, catch location, and discards. Instead, a program of mandatory vessel logbooks and dealer reports was implemented. Vessel trip reports for 1994, 1995, and 1996 were available for these stock assessments, but all data are provisional. Vessel trip report data include information on area fished and on retained and discarded catch and effort. There is no direct link between dealer report data and the vessel trip report, since there is no unique identifier that associates a vessel trip with a dealer transaction. Fields common to both data sets were used as an indirect link and provide some information on landings by market category and stock area (area fished). Observations that had zeros in either data set were eliminated to avoid erroneous matches. Because it was uncertain whether live or landed weight was given in a vessel report, the dealer and vessel trip reports were retained in matching sets to assist in determining the precision of allocations of species catches to stock groups.

Effort is estimated from the logbook and interview data. Catch per unit effort (CPUE; i.e., total pounds landed per days absent) is estimated from commercial weigh-out and logbook data. There is undoubtedly a difference in the effort recorded by the earlier method that used port agents' interviews and the current method that uses self-reporting by harvesters. Discard and effort information were evaluated by the SARC, and the results were compared to information obtained from corresponding observer data. The Northeast groundfish stocks reviewed are not managed by a TAC, but inaccuracies in reported removals should be expected nevertheless.

Trawl Surveys

Surveys have been conducted in the autumn since 1963 and in the spring since 1968 by NEFSC. The region is separated into 65 strata (Figure 1.6), and trawls are conducted at sites randomly chosen in each area. Beginning in 1985, trawl doors were changed from wood and steel to all steel. Conversion factors are used to adjust catches to make them comparable. In addition, surveys conducted by Canada are used in the stock assessment for Georges Bank cod and haddock. Data from scallop surveys are also used in stock assessments on southern New England yellowtail flounder. Recreational data are evaluated, but not used in the final VPA assessments of cod.

Age Sampling

Age composition is estimated by market category from length frequency and age samples. Data were pooled by calendar quarter, except that when samples were insufficient within a quarter, they were pooled semiannually. Most samples were obtained during port sampling. However, for Georges Bank yellowtail flounder and haddock, port samplings were low and were supplemented by uncategorized (no market category) samples collected during sea sampling conducted by NMFS.

Models Used

In recent years, both Canadian and U.S. managers have used the adaptive framework or age-structured model (ADAPT) (Conser and Powers, 1990; Gavaris and VanEeckhaute, 1996). The advantage of the ADAPT model over previously used virtual population analyses (VPA) and sequential population analysis (SPA) models is that it permits the accommodation of other data and can weight data sources based on their variability, reliability, and relative importance (Conser and Powers, 1990). The ADAPT models used by CAFSAC and NMFS differ in some of the equations used for weighting the reliability of ageing data and the incorporation of other observations, but they are essentially similar implementations (Conser and Powers, 1990).

Results

The results of the stock assessments were summarized in the advisory report on stock status (Table 1.3). Some of their conclusions are that (1) fishing mortality for cod, haddock, and yellowtail flounder in all areas except the Gulf of Maine has been reduced below the level of overfishing and is near or below the levels for rebuilding established by the fishery management plan; (2) recruitment is low relative to historical levels; (3) the spawning stock biomass has shown some rebuilding in all stocks except Gulf of Maine cod; and (4) except for Gulf of Maine cod, short-term projections indicate that spawning stock biomass will be maintained or rebuilt. Projections for Gulf of Maine cod indicate a continued decline.

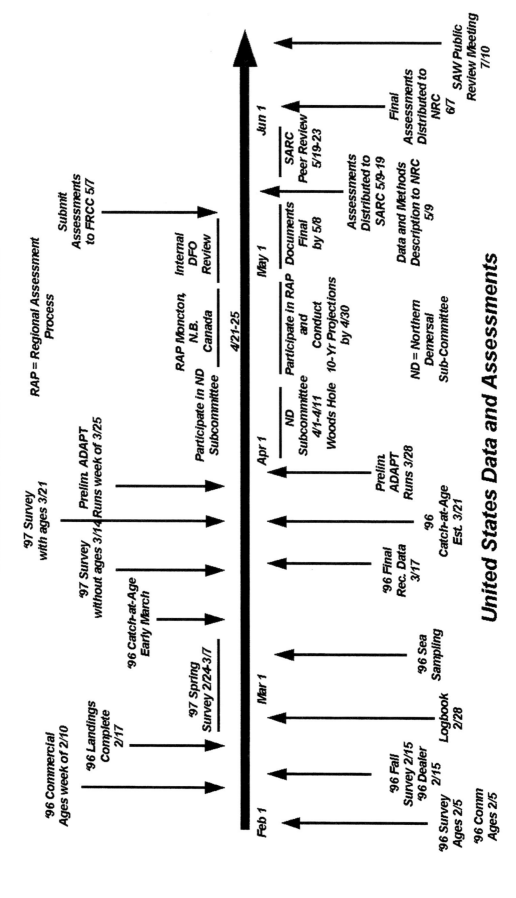

FIGURE 1.5 Time line for United States and Canadian data analysis and stock assessments in 1997. NOTE: ND = Northern Demersal, NRC = National Research Council.

Introduction

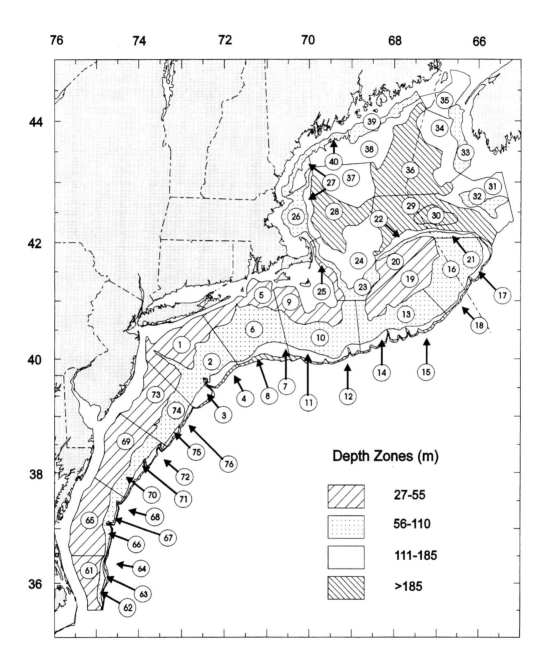

FIGURE 1.6 Map showing the sampling strata for NEFSC bottom trawl survey. SOURCE: NEFSC, 1997a.

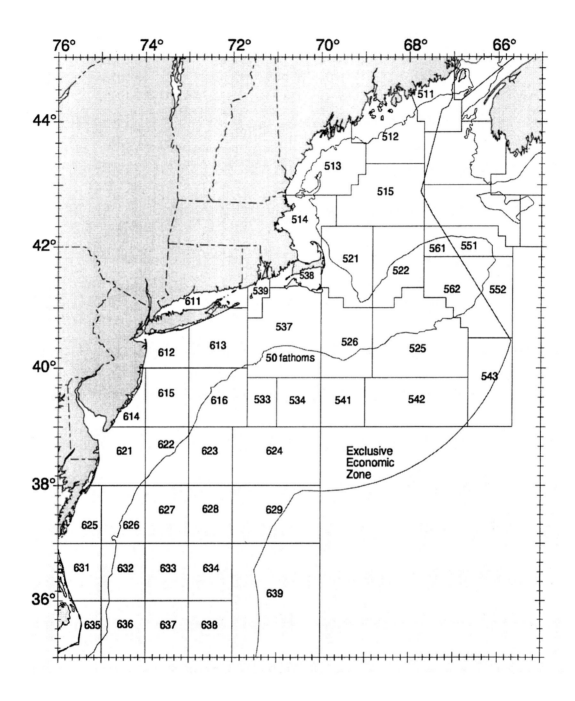

FIGURE 1.7 Map showing statistical unit areas off Northeast U.S. and Canada. SOURCE: NEFSC, 1997a.

TABLE 1.3   Status of Five New England Groundfish Stocks Reviewed at SAW-24

Stock	Current Fishing Mortality	1994-1996 Recruitment	1994-1996 Spawning Stock Biomass	Biomass Threshold
Gulf of Maine cod	Well above target	Low	Low declining	Threshold not well defined
Georges Bank cod	Near target	Low	Low increasing	Well below
Georges Bank haddock	Below target	Low	Low increasing	Below
Georges Bank yellowtail flounder	Below target	Average	Low increasing	Near
Southern New England yellowtail flounder	Below target	Low	Low increasing	Below

Status of each stock is summarized in terms of current (1996) fishing mortality, 1994-1996 recruitment, 1994-1996 spawning stock biomass (SSB), and minimum SSB thresholds established for management purposes in the multispecies FMP. Fishing mortality in 1996 is characterized relative to the $F_{0.1}$ or F_{max} (Gulf of Maine cod) rebuilding targets. Minimum SSB thresholds were established for Georges Bank cod (70,000 metric tons), Georges Bank haddock, (80,000 metric tons), Georges Bank yellowtail flounder (10,000 metric tons), and southern New England yellowtail flounder (10,000 metric tons) from Table 2 of NEFSC, (1997b).

HISTORY AND PURPOSE OF THIS STUDY

In the most recent reauthorization of the Magnuson-Stevens Fishery Conservation and Management Act (P.L. 104-293, 16 U.S.C. 1801 et seq.), Congress mandated that the National Academy of Sciences **"shall conduct a peer review of Canadian and U.S. stock assessments, information collection methodologies, biological assumptions and projections, and other relevant scientific information used as the basis for conservation and management in the Northeast multispecies fishery"** (see Appendix A). The National Research Council (NRC) has been charged with the review and has formed a committee to fulfill this mandate (see Appendix B for biographies of committee members). This report is a review of the 1997 stock assessments of cod, haddock, and yellowtail flounder conducted by NMFS. The committee also examined the stock assessments conducted by Canada for cod and haddock.

The committee's approach was to (1) attempt to replicate NMFS stock assessments; (2) compare the current stock assessment with other assessments around the world; (3) evaluate the stock assessment of cod, haddock, and yellowtail flounder according to the guidelines outlined in the *Improving Fish*

Stock Assessments (NRC, 1998); (4) evaluate the status of stocks of cod, haddock, and yellowtail flounder; and (5) evaluate the role of stock assessment in the fishery management process. In Chapter 2, the committee reviews the Northeast groundfish fish stock assessments in general to address whether they provide a valid scientific basis for decisionmaking. Included in this chapter is the committee's attempt to provide a more realistic depiction of uncertainty in forecasts of stock condition. In Chapter 3, a detailed review of the 1997 NMFS stock assessments in particular is provided to show where improvements could be made. An appraisal of the role of stock assessment in the management process is outlined in Chapter 4. The committee's findings and conclusions are summarized in Chapter 5.

NMFS provided background material and the 1997 stock assessments both before and after these had undergone its own review process. NMFS personnel made a presentation of the stock assessment to the committee. Personnel from the Canadian Department of Fisheries and Oceans also provided background material and stock assessments and presented the stock assessments to the committee.

The NRC retained a consultant, Marine Resources Assessment Group (MRAG) America, Inc. to assist the committee in its task. The committee requested that the consultant replicate analyses conducted by NMFS and carry out an alternative set of projections. These results were used to judge the quality of the stock assessment conducted by NMFS. In addition, the consultant was asked to extend the stock assessments by using data from more years than are typically used by NMFS. The results of the consultant's work were considered by the committee, and some results appear in this report.

The committee met in July 1997 in Bedford and Gloucester, Massachusetts. During this meeting, the committee heard presentations from invited speakers, other interested parties, and stakeholders to understand the context in which stock assessments are made.

2

GENERAL REVIEW OF NORTHEAST GROUNDFISH STOCK ASSESSMENTS

But that dread of something after death, the undiscover'd country from whose bourn no traveler returns, puzzles the will and makes us rather bear the ills we have than fly to others that we know not of.

William Shakespeare, *Hamlet*

Stock assessment is the science of data collection, analysis, and modeling that provides the basis for prudent, sustainable exploitation of fishery resources. It includes the provision of scientific advice about management strategies used to exploit fish stocks and the integration of science and scientific advice into the management process. In particular, the feedback between stock assessment and fisheries management has to be included to manage fisheries effectively.

A recent National Research Council (NRC) report *Improving Fish Stock Assessments* (NRC, 1998) reviewed the state of existing knowledge about stock assessment and made ten recommendations to improve the process (Box 2.1). The first recommendation is that a complete stock assessment should include five major topics: stock definition; data; assessment model; policy evaluation; and communication of results to managers and stakeholders. A checklist of items that should be in a stock assessment is given in Box 2.2. These recommendations provide a benchmark against which fishery stock assessments can be measured. The committee considered the framework of recommendations from the earlier NRC report presented in Boxes 2.1 and 2.2.

The approach of this committee was to examine the Northeast groundfish stock assessments against well-defined standards of quality. This investigation was a multistage process:

1. The committee first examined the data collection protocols and assessment models used. In particular, the following issues were evaluated: whether appropriate data were collected; whether stock assessments could be replicated; whether alternative models were used or should have been used; and whether forecasts of future populations were appropriate. In this chapter, the general results of this examination are presented, which are of interest to a general audience. Recommendations regarding technical details of stock assessments of interest to specialists are given in Chapter 3.

2. The committee then compared the assessments against approaches used around the world. The idea behind this comparison was to determine whether other stock assessment processes were qualitatively better than those for the Northeast fishery.

3. The committee evaluated the Northeast groundfish stock assessments against the NRC (1998) recommendations for improving stock assessment. Given that the earlier report was in press and had not been seen by stock assessment scientists, ours was a particularly severe test of the Northeast fishery assessments, which did not have the NRC guidelines.

Box 2.1: Recommendations of the NRC 1998 Report, Improving Fish Stock Assessments

1. Stock assessment scientists should conduct complete assessments using a checklist such as given in Box 2.2. Scientists from state and federal governments and from independent fisheries commissions should continue to conduct fish stock assessments with periodic peer review.

2. At the minimum, at least one reliable abundance index should be available for each stock. Fishery-independent surveys offer the best choice for achieving a reliable index if designed well with respect to location, timing, sampling gear, and other statistical survey design considerations.

3. Because there are often problems with the data used in assessments, a variety of different assessment models should be applied to the same data; new methods may have to be developed to evaluate the results of such procedures. The different views provided by different models should improve the quality of assessment results. Greater attention should also be devoted to including independent estimates of natural mortality in assessment models.

4. The committee recommended that fish stock assessments include realistic measures of the uncertainty in the output variables whenever feasible. Although a simple model can be a useful management tool, more complex models are needed to better quantify all the unknown aspects of the system and to address the long-term consequences of specific decision rules adequately. Implementation of this recommendation could follow the methods discussed in Chapter 3 (of the earlier report.)

5. Precautionary management procedures should include management tools specific to the species managed, such as threshold biomass levels, size limits, gear restrictions, and area closures (for sedentary species).

6. Assessment methods and harvesting strategies have to be evaluated simultaneously to determine their ability to achieve management goals. Ideally, this involves implementing them both in simulations of future stock trajectories. For complex assessment methods, this may prove very computationally intensive, and an alternative is to simulate only the decision rule while making realistic assumptions about the uncertainty of future assessments. Simulation models should be realistic and should encompass a wide range of possible stock responses to management and natural fluctuations consistent with historical experience. The performance of alternative methods and decision rules should be evaluated using several criteria, including the distribution of yield and the probabilities of exceeding management thresholds.

7. NMFS (National Marine Fisheries Service) and other bodies responsible for fishery management should support the development of new techniques for stock assessment that are robust to incomplete, ambiguous, and variable data and to the effects of environmental fluctuations on fisheries.

8. The committee recommended that NMFS conduct (at reasonable intervals) in-depth, independent peer review of its fishery management methods to include (1) the survey sampling methods used in the collection of fishery and fishery-independent data, (2) stock assessment procedures, and (3) management and risk management strategies.

9. The committee recommended that a standardized and formalized data collection protocol be established for commercial fisheries data nationwide. The committee further recommends that a complete review of methods for collection of data from commercial fisheries be conducted by an independent panel of experts.

10. NMFS and other bodies that conduct fish stock assessments should ensure a steady supply of well-trained stock assessment scientists to conduct actual assessments and carry out associated research. NMFS should encourage partnerships among universities, government laboratories, and industry for their mutual benefit. This can be accomplished by exchanging personnel and ideas and by providing funding for continuing education at the graduate, postdoctoral, and professional levels, including elements such as cooperative research projects and specialized courses, workshops, and symposia.

SOURCE: NRC, 1998.

Box 2.2 Checklist for Conducting or Reviewing Stock Assessments

Step	Important Considerations

1.0 Stock Definition
 	
    Stock Structure
    Single or multispecies
	What is spatial definition of a "stock"?
	Should the assessment be spatially structured or assumed to be spatially homogeneous?
	Choose single-species or multispecies assessment?
	Use tagging, microconstituents, genetics, and/or morphometrics to define stock structure?

2.0 Data	

2.1 Removals
 	
    Catch
    Discarding
    Fishing-induced mortality
	
	Are removals included in the assessment?
	Are biases and sampling design documented?

2.2 Indices of abundance	For all indices, consider whether an index is absolute or relative, sampling design, standardization, linearity between index and population abundance, what portion of stock is indexed (spawning stock, vulnerable biomass).
Catch per unit effort (CPUE)	What portions of fleet should be included and how should data be standardized? How are zero catches treated? What assumptions are made about abundance in areas not fished? Spatial mapping of CPUE is especially informative.
Gear surveys (trawl, longline, pot)	Is gear saturation a problem? Does survey design cover the entire range of the stock? How is gear selectivity assessed?
Acoustic surveys	Validate species mix and target strength.
Egg surveys	Estimate egg mortality, towpath of nets, and fecundity of females.
Line transect, strip counting	

2.3 Age, size, and sex-structure information
    Catch at age
    Weight at age
    Maturity at age
    Size at age
    Age-specific reproductive information
	Consider sample design, sample size, high-grading selectivity, and ageing errors.

Box 2.2—continued

2.4 Tagging data	Consider both tag loss and shedding and tag return rates. Was population uniformly tagged or were samples recovered?
2.5 Environmental data	How should such data be used in the assessment? What are the dangers of searching databases for correlates?
2.6 Fishery information	Are people familiar with the fishery, who have spent time on fishing boats, consulted and involved in discussions of the value of different data sources?

3.0 Assessment Model

3.1 Age-, size-, length-, or sex-structured model?	Are alternative structures considered?
3.2 Spatially explicit or not?	
3.3 Key model parameters	

Natural mortality
   Vulnerability
   Fishing mortality
   Catchability
	Are these parameters assumed to be constant or are they estimated? If they are estimated, are prior distributions assumed? Are they assumed to be time invariant?

Recruitment	Is a relationship between spawning stock and recruitment assumed? If so, what variance is allowed? Is depensation considered as a possibility? Are environmentally driven reductions (or increases) in recruitment considered?
3.4 Statistical formulation	

What process errors?
   What observation errors?
   What likelihood distributions?
	If the model is in the form of weighted sum of squares, how are terms weighted? If the model is in the form of maximum likelihood, are variances estimated or assumed known?

3.5 Evaluation of uncertainty	

Asymptotic estimates of variance
   Likelihood profile
   Bootstrapping
   Bayes posteriors
	How is uncertainty in model parameters or between alternative models calculated? What is actually presented, a distribution or only confidence bounds?

3.6 Retrospective evaluation	Are retrospective patterns evaluated and presented?

Box 2.2—continued

4.0 Policy Evaluation

    4.1 Alternative hypotheses

What alternatives are considered: parameters for a single model or different structural models?
How are the alternative hypotheses weighted?
What assumptions are used regarding future recruitment, environmental changes, stochasticity, and other factors?

Is the relationship between spawners and recruits considered? If so, do future projections include autocorrelation and depensation?

    4.2 Alternative actions

What alternative harvest strategies are considered?
What tactics are assumed to be used in implementation?
How do future actions reflect potential changes in future population size?
Is implementation error considered?
Are errors autocorrelated?
How does implementation error relate to uncertainty in the assessment model?

    4.3 Performance indicators

What is the real "objective" of the fishery? What are the best indicators of performance? What is the time frame for biological, social, and economic indices? How is "risk" measured? Are standardized reference points appropriate? Has overfishing been defined formally?

5.0 Presentation of Results

How are uncertainties in parameters and model structure presented? Can decision tables be used to summarize uncertainty and consequences? Is there explicit consideration of the trade-off between different performance indicators?

Do the decisionmakers have a good understanding of the real uncertainty in the assessment and the trade-offs involved in making a policy choice?

SOURCE: NRC, 1998.

4. Finally, the committee interpreted the stock assessments to determine the appropriate scientific advice to be drawn from them. In particular, it determined stock size and condition, whether exploitation rates were high, whether current regulations have reduced fishing mortality, whether lower fishing mortality would diminish the yield obtained from a fixed amount of recruitment, and whether there appeared to be a relationship between recruitment and spawning biomass or between recruitment and fishing mortality. These factors are routinely examined in providing scientific advice about acceptable catch levels, as explained below.

INVESTIGATION OF NORTHEAST FISHERY STOCK ASSESSMENTS

For the most part, the Stock Assessment Review Committee (SARC) and auxiliary reports contained information about the five major topics in stock assessment (Box 2.2): stock definition; data; assessment model; policy evaluation; and presentation of results to managers. The more detailed recommendations follow:

- **Stock definition:** Stock identification issues have been considered in the stock assessments, leading to independent assessments for Georges Bank cod, haddock, and yellowtail flounder; Gulf of Maine cod; and southern New England yellowtail flounder; as well as for some 50 other stocks in the general area. It should be noted that the stock boundaries for the U.S. and Canadian assessments are different because the Canadians estimate the biomass of fish inhabiting their waters and set a total allowable catch (TAC) as a proportion of this biomass. U.S. assessments cover the stock in both U.S. and Canadian waters to provide information on total stock. As a consequence, the information in U.S. and Canadian assessments is complementary, not contradictory, and scientists from each country participate in both assessments. The committee notes that better information on genetics and migrating behavior of these populations is needed in order to establish causal mechanisms for changes in stock size by area

- **Data:** The National Marine Fisheries Service (NMFS) assessments contain information and documentation of a variety of data, including landings, discards, logbook information, age and other biological sampling information, survey indices of abundance, and various analyses of the data to provide standardization. There are problems with aspects of data collection (see Chapter 3), but these problems are identified (see Appendix D). It was beyond the scope of the committee to conduct an in-depth review of the raw data.

- **Assessment model:** The ADAPT assessment model used in the assessments is documented, and in the case of Gulf of Maine cod, an alternative analysis was conducted using concepts developed by Fournier and Archibald (1982) and others (Ianelli, 1997; see NRC, 1998 for additional information). The main source of variability considered is in survey indices of abundance. The committee accepted the use of ADAPT as the primary assessment model but provided comments on its features and alternatives in Chapter 3. Methodology for projecting the future population under alternative scenarios is documented and allows for variation in recruitment and starting abundance. The committee examined spawner-recruit data from the five stocks (see the section "Status of the Five Stocks", (pp. 40-60) and Appendix F) and noted that a wide variety of models could be fitted to the data and that alternative interpretations of what will happen with future recruitment are valid.

- **Policy evaluation:** Alternative hypotheses for policy evaluation are considered to a limited extent, mainly in the pattern of expected recruitment. Alternative actions considered include a range of different fishing mortalities (Fs) without consideration of implementation error (deviations from fishing mortalities due to the dynamic nature of fishing and regulation). The main performance indicator is the statistical distribution of spawning biomass over a 10-year period, which is condensed into a risk measure related to the probability of reaching a rebuilding target set by the Northeast Fishery Management

Council (NEFMC). The committee is concerned that uncertainty tends to be underestimated in these projections, with the consequence that probabilities of not reaching the rebuilding targets may be higher than suggested. Consequently, in "Evaluating the Consequences of Alternative Management Actions," the committee gives a rationale for considering a wider range of alternatives and illustrates this approach by reexamining the Georges Bank haddock assessment.

- **Presentation of results**: Results are presented in a concise advisory report, an in-depth report of the Stock Assessment Workshop (SAW), and related technical documents and reports (see Appendix C for a list of material provided to the committee). Improvements could be made in all of these areas, but it is clear that considerable documentation and analysis are available.

Replication of Assessment Results

The committee's consultant, Marine Resources Assessment Group (MRAG) Americas Inc., reran the ADAPT models, using workspaces provided by NMFS and was able to replicate the NMFS stock assessment results. The committee did not have the resources to investigate raw data sources, so it utilized NMFS data summaries. In addition, the consultant used a different ADAPT program, frequently employed in Canadian assessments (Gavaris, 1991; Gavaris et al., 1996, Gavaris and VanEeckhaute, 1997), that treats the oldest age group somewhat differently. The results were very similar to the NMFS assessments, with average annual differences in age 1 and total abundance (numbers of fish) being less than 3% for all five stocks.

Evaluating Consequences of Alternative Management Actions

A primary objective of fish stock assessments is to evaluate the possible consequences of alternative management decisions. This evaluation is accomplished by simulating future stock projections under different management options and assessing gains and risks associated with each. The ability to predict future stock responses to management interventions is, in general, very limited, and this fact should be reflected in the simulations. Uncertainty about future stock projections has several sources: (1) uncertainty about the current status of the stock (*assessment error*); (2) variability of future stock dynamics, such as environmental effects on recruitment (*process error*); (3) uncertainty about how to model the future dynamics of the population, including recruitment, growth, mortality and other relevant processes (*model uncertainty*); and (4) errors in the implementation of management strategies (*implementation error*).

Errors in the implementation of management strategies only compound the inherent biological uncertainties. Managers can only attempt to control fishing mortality indirectly through effort and/or catch regulations or, more directly, by closing grounds to fishing. In either case, the link between regulatory tactics and the resulting fishing mortality is uncertain. One of the problems is that it is difficult to predict how the fishing industry will respond to a given set of regulations. Delays or adjustments in management strategies in the Northeast due to the industry's response have been typical. As discussed in Chapter 4, these problems can sometimes be reduced when stakeholders are involved in developing management strategies early in the process.

Assessments of the Northeast groundfish stocks include an evaluation of the consequences of setting different fishing mortality over a 10-year period (see Appendix D and NEFSC, 1997a). Among the management alternatives considered for the five stocks were to maintain the current fishing mortality and to implement a mortality of $F_{0.1}$. In addition, the effects of closing down the fishery were evaluated for Gulf of Maine cod, and $F = 0.1$ was explored for Georges Bank haddock. Overall, the committee believes that these projections have underestimated the uncertainty inherent in predicting future stock responses to different management regulations.

NMFS scientists did incorporate uncertainty about the current stock status in simulations. However, as is standard practice in fishery stock assessments, the level of uncertainty was evaluated on the assumption that the assessment model was correct. Especially problematic in the case of Northeast groundfish assessments are the assumptions that natural mortality is fixed and known and that catches-at-age are observed without error. These assumptions about mortality and catch-at-age result in overly optimistic assessments of possible estimation errors and, in turn, an underestimation of the variability of possible stock sizes at the start of the simulations (see Chapter 3). An alternative model that incorporates these two factors (Ianelli, 1997) showed a greater range of uncertainty, and further efforts of this type are desirable.

With respect to uncertainty in the stock dynamics, a wide range of possible stock responses is usually consistent with historical experience and should be considered in the simulations. In many situations (e.g., in the Northeast groundfish complex), a discrete set of alternative scenarios can be identified to characterize possible future trends in recruitment. The projections conducted by NMFS are instead based on a single, "best-fit" stock-recruitment model. A Beverton-Holt model was fitted to the time series of estimates of spawning biomass and recruitment provided by the assessment model, with assessment errors ignored. Residual variability in the stock-recruitment process was incorporated into the simulations, but no uncertainty in the specification of the model was considered.

In general, the results of these projections indicate that stocks, recruitment, and future catches will increase if fishing mortalities are reduced substantially relative to recent high values. Although, in most cases, stock and recruitment time series indicate low recruitment levels on average in recent years when the stocks were depleted, there is no guarantee that these trends would reverse if the stocks recovered. The possibility that historical trends in recruitment were driven by changes in the ecosystem cannot be ruled out. So the alternative that recruitment may not recover to historical high levels when and if stocks rebuild should be considered. Also, an evaluation of risks under high fishing mortality hinges on how future recruitment may be affected if the stock is kept at very low levels. Some stock-recruitment (*S-R*) plots appear to be consistent with a *depensatory* relationship, in which the number of recruits produced per unit of spawning biomass decreases as the stock becomes more severely depleted. This depensation alternative should be considered in such cases. An interesting approach that may be used to postulate alternative hypotheses about the relationship between spawning biomass and subsequent recruitment is to compare stock-recruitment patterns across populations of a single species or groups of similar species. These comparisons may be used to evaluate the possibility of depensation (e.g., Liermann and Hilborn, 1997; Myers et al., 1995), the capacity of the stock to recover from low abundance levels (e.g., Myers et al., 1997), or simply to see how parameters estimated for a particular stock fit in relation to other similar stocks.

To illustrate how future stock responses to management can change for different stock-recruitment scenarios, the committee conducted a limited set of simulations using Gulf of Maine cod, the most problematic stock, which is still subject to high levels of fishing mortality. Four alternative recruitment scenarios were postulated based on assessment results: (1) recruitment will increase in proportion to stock size if spawning biomass is allowed to increase; (2) recruitment will stay constant on average at the historical mean value independent of stock size; (3) the stock-recruitment relationship shows depensation at very low stock size; and (4) the same stock-recruitment model used in NMFS assessments. In all cases, residual variance was assumed to be uncorrelated from year to year. Further details about the methods and results are provided in the section "Status of the Five Stocks" (pp. 40-60) and Appendix F.

The Beverton-Holt recruitment function did not fit the data particularly well for Gulf of Maine cod, as well as some other stocks, which motivated the use of the depensatory model. Whether the lack of fit in the recruitment data were due to variability related to environmental conditions or to depensation in the spawner-recruitment relationship cannot easily be resolved. In the NMFS analysis, the age at maturity for George Bank cod is assumed to be stable throughout the past 10 years, whereas for Gulf of Maine cod, age at maturity has increased. This type of population response at low stock sizes is contrary to what could be expected: if the carrying capacity of the stocks is constant, one would rather expect a decrease in age at maturity when stocks decline. If the assumed increase in age at maturity were artificial, the spawning biomass would be underestimated in recent years, and the spawner-recruit relationship would have even stronger depensation than estimated. Other hypotheses related to carrying capacity changes also are possible. Understanding possible processes affecting recruitment at small stock sizes would require a more refined formulation of the spawning process, where the sex and age structure of the stock are taken into account as they affect egg production.

The simulations show that reducing fishing mortality to the $F_{0.1}$ level resulted in increases in stock size in all recruitment scenarios (see the section "Status of the Five Stocks" (pp. 40-60) and Appendix F). Increases were only moderately larger when recruitment was assumed to increase in proportion to spawning biomass than when recruitment was independent of stock size. This occurred because most of the increase in adult biomass is due to a reduction in mortality of recruits, and very little is due to increases in recruitment. Larger gains derived from improved recruitment would be realized later if this scenario is correct. Maintaining fishing mortalities at the current high levels, on the other hand, would have very contrasting effects depending on the stock-recruitment scenario. On one extreme, under the depensatory stock-recruitment relationship, the stock continued to decline and collapsed in all trials after six to nine years. Similar results, although the stocks did not become extinct, were obtained when recruitment was proportional to stock size. On the other extreme, when recruitment was assumed to be independent of spawning biomass, predicted stock size increased only slightly, since average recruitment was set equal to the average of the last 15 years, which is somewhat higher than the most recent recruitment estimates.

Results of this type can be summarized in a decision-analysis table in which the consequences of alternative actions can be evaluated across different recruitment scenarios (Table 2.1). To be even more useful in decisionmaking, such tables should be constructed using actual management tactics that could be employed to implement different target fishing mortalities, rather than using target fishing mortalities themselves.

How conservative management actions should be depends on the probabilities assigned to alternative scenarios. For example, assigning a high probability to the depensatory model would prompt severe restrictions in fishing effort to minimize the possibility of stock collapse. On the other hand, if depensation was considered unlikely and recruitment was assumed to be driven mostly by the environmental conditions, the motivations to rebuild stocks rapidly would not be as strong. In many cases, assigning probabilities is difficult when views contrast on how nature may operate based solely on the stock-recruitment data. Independent observations of similar stocks could, for instance, be used to assess how likely depensation may be (Myers et al., 1995). Also, studies addressing possible links between recruitment and environmental changes may provide some evidence for or against hypotheses involving environmental change. Ecosystem changes have been studied extensively in the area, and these studies could have a more prominent role in assessment, especially in the construction of alternative recruitment scenarios.

Because of the underestimation of uncertainty and the limited set of management options explored, the analysis presented cannot be used to evaluate trade-offs between possible gains and risks under more or less stringent management regulations. The committee recommends that a more comprehensive analysis of the effects of alternative management options incorporate the three main sources of uncertainty discussed above. In addition, socioeconomic factors may be included and would also be subject to similar peer review as the stock assessments.

TABLE 2.1 Consequences of Implementing Different Rates of Fishing Mortality Under Alternative Stock-Recruitment (*S-R*) Scenarios for Gulf of Maine Cod.

Recruitment (*R*) scenarios	Fishing Mortality		
	$F = 0.16$ ($F_{0.1}$)	$F = 0.29$ (F_{max})	$F = 1.04$ (10-year mean)
R proportional to stock size	SSB increases by 274 to 1488% Catches increase by 247 to 1315%	SSB increases by 107 to 850% Catches increase by 85 to 801%	SSB continues to decline -66 to -92% Substantial drop in catches -67 to -92
R independent of stock size	SSB increases by 306 to 1240% Catches increase by 288 to 1214%	SSB increases by 175 to 867% Catches increase by 154 to 840%	Slight increase in SSB Slight increase in catches
Depensatory *S-R* relationship	SSB increases by 365 to 1503% Catches increase by 329 to 1450%	SSB increases by 14 to 568% Catches increase by 2 to 534%	Stock collapses with probability = 1
S-R model used in NMFS projections	SSB increases by 307 to 1204% Catches increase by 272 to 1158%	SSB increases by 142 to 760% Catches increase by 122 to 718%	SSB declines 0 to -88% Catches decline -11 to -89%

NOTE: Results are percent change in spawning stock biomass (SSB) and catch at the end of a 10-year projection. Range corresponds to 2.5 and 97.5 percentiles of 1000 trials as a percentage of the median at the start of projections.

COMPARISON WITH ASSESSMENTS AROUND THE WORLD

Input Data

In the Northeast, as is common worldwide, assessments are based on age-structured assessment methods, in this case using a particular age-structured model calibrated to time series of survey catch rates. The data necessary to conduct such analyses are available in the Northeast and are comparable to the data routinely used in stock assessments elsewhere, although Northeast data quantity and quality could be improved. It would be particularly useful to collect additional biological samples, to increase the number of sets made during the surveys, and to improve the reliability of catch-and-effort data, as explained in Chapter 3. Inaccurate landing statistics are a widespread problem in stock assessments around the world that can be particularly acute in TAC-managed fisheries, where the incentives to misreport catches are greater.

Method and Calibration

Several methods can be used in stock assessments that utilize catch-at-age information, and particular methods tend to be associated with specific geographic areas. For example, stocks assessed by the International Council for Exploration of the Sea (ICES) most often are analyzed by the Extended Survivor Analysis method (Anonymous, 1992). Those covered by the International Commission for the Conservation of Atlantic Tunas (ICCAT) are assessed with the ADAPT methodology (Gavaris,

1993), whereas stocks in the northeast Pacific are assessed with statistical models developed for several data sources (e.g., Stock Synthesis [Methot, 1989]; CAGEAN [Deriso et al., 1985]). Some of these methods have been evaluated (Patterson and Kirkwood, 1993; NRC, 1998), and the results indicate that their performance is generally comparable. Northeast stock assessments use ADAPT in a standard formulation where stock size is estimated by minimizing the square of the difference between the natural log of predicted stock size index (indices) minus the observed stock size index. Northeast stock assessments use ADAPT in a standard formulation where stock size is estimated by minimizing the sum of squared differences between the natural logarithms of predicted and observed stock size indices.

Analyses Included in the Assessment

The Northeast groundfish stock assessments include all the analyses expected by the committee: regression analysis to standardize catch rates (when CPUE [catch per unit effort] data are available and considered useful), ADAPT analysis to estimate stock size yield, spawner-per-recruit analyses, and stock-recruitment analyses to estimate biological reference points. Short-term projections are made with recruitment estimates when available, and stochastic medium-term projections are made to compare the effects of various fishing mortality rates over a 10-year period (see discussion of assessment models in Chapter 3). Uncertainties are likely greater than indicated in the assessment, but this is a problem shared by most other assessments based on virtual population analysis (VPA). The medium-term projections presented are all based on a Beverton-Holt stock-recruitment model, constrained in some cases so that the number of recruits per unit of spawning biomass did not exceed the median observed value when spawning biomass dropped below the historical minimum. No alternative recruitment scenarios were explored. In addition, the primary management measures used, days at sea and closed areas, are not explicitly taken into account in making the projections.

Provision of Advice

The provision of advice involves the collection of data, data analysis, documentation, peer review of the analyses formulation, and communication of advice. Ways in which these steps are performed around the world vary.

Methods for Regional Stock Assessment

U.S. Northeast Coast

On the U.S. Northeast coast, these steps are performed by SARC, implemented in the region in 1985. The assessments under review are conducted collaboratively by NMFS and Canadian Department of Fisheries and Oceans (DFO) analysts and are peer reviewed in both Canada and the United States. In the United States, a peer review is provided first at working group meetings by federal, state, and academic scientists (see Figure 1.5). A second peer review takes place at the Stock Assessment Workshop (SAW) where draft advice is formulated. The assessments and advice are then presented at a SAW plenary, where the final advice is formulated.

Canadian Maritimes

In the Canadian Maritimes Region, multidisciplinary stock assessment teams, including assessment scientists, oceanographers, and in some cases, individuals from the fishing industry, prepare an assessment. Preparation of the assessment may involve several meetings of each assessment team.

These assessments go through a first review generally within each laboratory (Halifax-Dartmouth, Moncton, St. Andrews). A second review takes place during the regional advisory process (RAP) involving federal, provincial, and academic scientists, and possibly fishery managers and harvesters or processors as well (see Figure 1.5). A consensus stock status report is drafted, agreed upon, and presented to the Fisheries Resource Conservation Council (FRCC), whose members, mostly from the fishing industry and academia, have been nominated by the Minister of Fisheries and Oceans. FRCC provides management advice to the Minister.

U. S. West Coast

On the U.S. West Coast, most assessments are done by NMFS analysts and then peer-reviewed by plan development teams (where active) composed of federal (including NMFS), state, and academic scientists. These assessments are then reviewed by Statistical and Scientific Committees (SSCs), also composed of federal, state, and academic scientists, which provide direct testimony to the regional fishery management councils. These councils frequently manage fish stocks with TAC levels, so that related scientific recommendations come in the form of acceptable biological catch (ABC) limits. The councils rarely (if ever) allow their recommended TACs to exceed the ABCs. In contrast, the Northeast fishery assessment advice is more qualitative, presumably because NEFMC does not practice TAC management. Nevertheless, the assessment review and recommendations in the Northeast fishery assessment are comparable to those from the West Coast.

Assessments and Management of Fish Stocks by International Council for the Exploration of the Sea (ICES)

The approach to assessments, advice, and implementation varies considerably within the ICES area. ICES has 19 member nations, conducts marine research in the North Atlantic ocean, and provides scientific advice to member countries. ICES coordinates fish stock assessments and advice through its advisory body, the Advisory Council on Fisheries Management (ACFM). This body in turn works as a review panel, formulating advice based on assessments and draft advice from area-based assessment working groups. In addition, ACFM coordinates the work of methodology-oriented working groups.

Membership on ACFM is mainly through national representation, in addition to some working group chairs and ex officio members. Membership in the working groups is by delegation from various member countries.

ACFM recommendations are passed to whatever party has asked for advice. This may be a country in the ICES area, the European Union, or a commission such as the Northeast Atlantic Fisheries Cooperative, North Atlantic Salmon Commission, or International Baltic Sea Fisheries Commission. Traditionally, this advice has been a response to questions about the size of the TAC for the coming year, but recently the emphasis has moved toward the evaluation of medium-term strategies, including elements of risk analysis.

A noticeable lack of dialogue is evident between advisors at various levels of the process and recipients of the advice. Thus, no forum exists for discussing appropriate management strategies, elucidating appropriate management targets, and so forth. The exchange of information is limited: the stock assessment along with annual TAC advice are given when requested. This situation has caused major problems in the past, ranging from criticism of ACFM for giving explicit advice when no danger signals are present, to severe criticism of ACFM for not providing advice until stocks are well into an overfished state. The present evolution into a process that emphasizes the medium-term consequences of different harvesting strategies is the result of an attempt to alleviate these problems.

Apart from concerns about its form, ACFM advice has generally been received without much criticism, or calls for re-evaluations. This is due, in part, to the international nature of the process and to the fact that traditionally all ACFM advice is given as consensus advice.

Assessments and Management of Whale Stocks by the International Whaling Commission (IWC)

The guidelines for stock assessment and management for commercially exploitable whale stocks are contained in the Revised Management Scheme, developed mainly by the Scientific Committee (but not adopted by the IWC). The Revised Management Scheme consists of a monitoring and observer scheme and a management procedure. The management procedure is implemented for a given fishery through an extensive computer-based risk analysis and consists of subdividing the ocean basin into subareas and specifying how the catch limit algorithm is to be applied to various subareas.

Catch limits are calculated from historical catches and past and current survey data by way of a Bayesian-like computation. Surveys are conducted regularly, and the abundance estimates are reviewed by the IWC Scientific Committee. These reviews are usually done every five years and if the Scientific Committee agrees that a set of abundance estimates is acceptable as input to the catch limit algorithm, catch limits for the coming five-year period are calculated. If acceptable abundance estimates are not available, the fishery is phased out over 10 years.

More comprehensive stock assessments are carried out when the management procedure is implemented for the stock and when a revised implementation is required (e.g., when area definitions change or alternative management measures are considered).

Compared to the stock assessment carried out under the IWC Revised Management Scheme (and actually undertaken only for Norwegian minke whaling in the northeast Atlantic), the assessment and management of the five groundfish stocks off New England are characterized by: (1) a higher frequency of assessments; (2) less external scrutiny of the assessments; and (3) no clear long-term management context for assessments. The last difference is the most striking. An implementation of the management procedure represents a long-term management strategy that has been put to extensive risk analysis for long-term behavior. For New England groundfish stocks, the assessment results have not been handled within the framework of an explicit long-term strategy. Therefore, no *formal* long-term risk analyses of future stock development with an emphasis on collapse and continuing yield have been undertaken. However, several economic analyses have been conducted for the region (Edwards and Murawski, 1993; Overholtz et al., 1993).

Precautionary Management Advice

Scientific advice to fisheries managers has become more precautionary as scientists have learned that sustainability of fish resources requires less intense harvesting or other stringent management measures. The extra precaution has become necessary due to uncertainty about the accuracy of assessment inputs and methods, and errors in implementation of management measures (see NRC, 1998, Chapter 4). For example, the most common biological reference point to use as a target in the 1960s and 1970s was the fishing mortality producing maximum sustainable yield (F_{MSY}). In the 1990s, F_{MSY} has become a limit reference point to be avoided with high probability (FAO, 1995). Other common reference points include fishing mortalities related to maximizing yield per recruit (F_{max}) and the point at which the marginal increase in yield per recruit is reduced to 10% ($F_{0.1}$). These also may not be suitably conservative for some stocks with variable recruitment, maturation at relatively old ages, or large implementation errors (Clark, 1993; Mace and Sissenwine, 1993). Since the late 1980s, biological reference points such as fishing mortalities that preserve spawning biomass per recruit at 35 to 45% of the unfished level (denoted $F_{35\%}$ for example) are becoming more frequently used, particularly on the west coast of the United States. More recently, techniques which further reduce fishing mortality at low spawning biomass are being considered to further minimize risk of stock collapse (NRC, 1998).

Table 2.2 shows recent estimates of fishing mortality from the stock assessments along with selected biological reference points. $F_{40\%}$ is selected to represent a typical fishing mortality with a low probability of stock collapse and $F_{20\%}$ is selected because it corresponds to the definition of overfishing used for many stocks by the NEFMC. Recent management measures by the NEFMC have reduced fishing mortality to or slightly below $F_{0.1}$ for four out of five stocks considered in this report, except Gulf of Maine cod. Furthermore, the values for $F_{0.1}$ are similar to those for $F_{40\%}$, suggesting that current estimated fishing mortalities are approaching the upper limit of what is considered conservative in the scientific community. Recent management measures by NEFMC have had no effect for Gulf of Maine cod, which remains fished at much above F_{max}, which itself is not sustainable in many situations (e.g., Deriso, 1982).

It should be emphasized, though, that scientific advice provided by NMFS and the SARC process has been in the spirit of the precautionary approach. They have recommended that fishing mortality not be allowed to increase for four of the five stocks and that strong management measures be used to sharply reduce fishing mortality for Gulf of Maine cod (NEFSC, 1997b). Nevertheless, for any of the five stocks, there are scientists who would recommend lower fishing mortalities than are currently estimated to occur. For example, the default target fishing mortality for groundfish managed by the North Pacific Fishery Management Council is $F_{40\%}$ scaled down by the ratio of current biomass to a target biomass level. For a stock at 1/2 of its target biomass, the default fishing mortality is $1/2\, F_{40\%}$. If this strategy were applied to Georges Bank haddock for example, the recommended fishing mortality would be roughly $1/2 \times 0.2 = 0.1$, since spawning biomass is at roughly 1/2 of the Council-recommended threshold.

Summary

Internationally, there are several models for the stock assessment process. The assessment process followed in Northeast fishery stock assessment and management appears open and involves steps analogous to those implemented for other fisheries. Critical evaluation of the assessments is undertaken, and the techniques used are comparable to those in other jurisdictions.

TABLE 2.2 Fishing Mortality (F) for the Five Northeast Groundfish Stocks, 1993-1996, $F_{0.1}$ and F_{max} a Yield-per-recruit curve, and $F_{20\%}$ and $F_{40\%}$ from a Spawning Biomass-per-recruit Curve.

Stock	$F(93)$	$F(94)$	$F(95)$	$F(96)$	$F_{20\%}$	F_{max}	$F_{0.1}$	$F_{40\%}$
GOM cod	0.9	2.1	1.1	1.0	0.4	0.3	0.15	0.15
GB cod	1.0	1.1	0.4	0.2	0.4	0.3	0.15	0.15
GB haddock	0.5	0.4	0.15	0.2	0.7	>1	0.25	0.2
GB yellowtail flounder	1.0	1.7	0.3	0.1	0.7	0.6	0.25	0.2
SNE yellowtail flounder	0.8	0.8	0.25	0.1	1.4	>1	0.3	0.3

SOURCE: NEFSC, 1997a. Fishing mortalities rounded to the nearest 0.05 between 0 and 0.3 and to the neareast 0.1 above 0.3. NOTE: GOM = Gulf of Maine; GB = Georges Bank; SNE = southern New England.

GENERAL EVALUATION OF NORTHEAST GROUNDFISH STOCK ASSESSMENTS BASED ON RECOMMENDATIONS FROM *IMPROVING FISH STOCK ASSESSMENTS*

Concise statements of the 10 recommendations (Box 2.1) from *Improving Fish Stock Assessments* (NRC, 1998) are italicized and then followed by a qualitative discussion of how well the Northeast fishery stock assessments compared.

1. *Stock assessments should contain the information identified in Box 2.2.* As described above, the Northeast fishery stock assessments contain most of this information.

2. *At least one reliable abundance index should be available for each stock.* In the Northeast fishery, indices of abundance come from the autumn and spring NMFS surveys and a Canadian spring survey. The amount of survey information is large compared to that available for many other places, and the reliability of the surveys has been ascertained. However, the amount of survey information is not large compared with information available for Canadian East Coast stocks and many stocks assessed in ICES (Pálsson et al., 1989; ICES, 1993). The committee finds that the level of effort for a particular survey may be too low (see Chapter 3).

3. *A variety of assessment models should be used, and independent estimates of mortality (M) should be considered.* Alternative assessment models have been used to only a limited degree. Little consideration has been given to independent estimates of mortality. Therefore, stock assessments are somewhat deficient in this regard.

4. *Stock assessments should include realistic measures of uncertainty in output variables, and more complex models should be considered to provide these measures.* The alternative assessment of Gulf of Maine cod (Ianelli, 1997) follows this approach. For the other assessments, attempts were made by NMFS to capture the uncertainty in models and projections. However, as described above, the committee believes that uncertainty is understated by NMFS in the present assessments.

5. *Precautionary management measures should include tools specific to the species managed.* Although this is not strictly an assessment issue, NMFS and SARC have provided scientific advice for many years specific to the species and area managed with regard to the effects of fishing mortality, size limits, and closed areas. Current management measures include a combination of tools that have apparently led to lower fishing mortality for four of the five stocks considered in this report.

6. *Assessment methods and harvesting strategies should be evaluated simultaneously to determine their ability to achieve management goals.* In NMFS assessments, this has been done by making the projections and looking at rebuilding probabilities. The major disconnect in the process is in determining how the management system, which is based on area closures and days at sea limitations, affects the fishing mortality of the stocks. The current assessments evaluate only the effects that different rates of fishing mortality would have on the probability of achieving rebuilding targets. Links between fishing mortality targets and management tactics were not analyzed. In other words, the implementation problem was not addressed in the assessments.

7. *New data and models should be developed that either can reduce uncertainty or are robust to incomplete, variable data or environmental fluctuations.* The committee found that NMFS scientists are considering new approaches in their stock assessments and are eager to respond to old problems and new challenges. However, NMFS should further advance the level of its science as suggested in this chapter and in Chapter 3. It was beyond the scope of this committee to conduct an in-depth review of the raw data. There are problems with aspects of data collection (see Chapter 3), but these problems are identified (see Appendix D). It may be worthwhile for NMFS to have an independent audit of the raw data undertaken.

8. *Periodic, independent peer reviews of assessments should be done.* The SARC process does provide for peer review and appears to be leading to constructive changes in assessment methodology,

data collection, and research. Nevertheless, SARC includes several NMFS personnel, so the review is not strictly independent. Thus, periodic outside review completely independent of NMFS should be implemented. Any regular assessment system and management advice will have a tendency to become more or less entrenched. Therefore, regular exchanges between various entities within the system and between systems are needed. Any well-designed system allows for the injection of new ideas and methods. External peer review should examine not only the stock assessment analyses but also the process of how stock assessment is linked to management concerns.

9. *Further documentation of standardized and formalized data collection systems is needed.* In the Northeast fishery assessment system, NMFS has recently documented data collection protocols in the catch, survey, and sampling areas (see Appendix C).

10. *NMFS should be encouraged to form partnerships with universities, government labs, and industry for the exchange of personnel and ideas and to provide funding for continuing education to keep the stock assessment process fresh and invigorating.* In the NMFS presentations to the committee and at the public hearing, there were indications that such activities are under way. Communications between NMFS and industry seem to be improving with sharing of personnel on research and commercial vessel operations. Such activities should be strengthened.

In summary, many of the earlier NRC report recommendations are already being addressed by NMFS even though it had not yet seen the report. The current stock assessment process, despite the need for improvements, clearly provides a valid scientific context for understanding fish populations and the effects of fishery management.

STATUS OF THE FIVE STOCKS

The final task of the committee was to interpret the assessments of the status of the five stocks based on information provided by NMFS and DFO and on analytical results from its consultant (discussed earlier; see Appendix F). In the early 1990s, NMFS stock assessments suggested that the five stocks had similar characteristics: low spawning-stock biomass (SSB) relative to 20 years ago and very high fishing mortality rates, with 50-80% of the fish being captured every year. These assessments advised that maintaining the high fishing mortality rates would lead to continued low catches and the likelihood of the total collapse of stocks and catch.

The need for major reductions in fishing pressure is predicated on the following assumptions:

1. Current stock size is low relative to recent history.
2. Recent fishing mortality has been high and is not sustainable.
3. Recent recruitments have been low relative to earlier periods, presumably because of low spawning-stock biomass, presence of depensation, and/or possibly other factors such as environmental changes.
4a. Reducing fishing pressure will allow spawning stocks to rebuild and/or recruitment to increase.
4b. Maintaining high fishing mortality rates has a high probability of leading to continued poor recruitment, making current yields non-sustainable.

Points 4a and 4b are the most important links in the chain. Although the recent high rates of fishing mortality are clearly incompatible with sustainable management, the extent of the biological benefits from reducing fishing pressure will depend on whether or not rebuilding spawning stocks results in higher recruitments.

The committee evaluated the evidence to support these assumptions. Most data and documentation can be found in Appendix D and/or NEFSC (1997a), and accompanying working papers. The committee examined plots of landings, CPUE, spawning-stock biomass, recruitment, and yield per recruit. Different spawner-recruit relationships for the five stocks were produced by the consultant to the committee and show data from ADAPT, fitted Beverton-Holt (B&H) curves produced by NMFS and the consultant, and curves fit using a density-independent model (Figures 2.1 through 2.5; see Appendix F for details).

The committee made a determination of whether a stock had collapsed by examining the historical estimates of spawning biomass and recruitment. If current estimates of spawning biomass were near the low end of historical values and if there had recently been little or no estimated recruitment for a number of years compared to historical estimates, then the committee stated that the stock had collapsed. This definition was used in mind of the Magnuson-Stevens Fishery Conservation and Management Act definition of "overfished" as being fished down to a level that jeopardizes the capacity of a stock to produce maximum sustainable yield on a continuing basis (16 U.S.C. 1801 et seq.). A stock at a low spawning level combined with low or little recruitment would be in such jeopardy.

To provide supporting evidence for trends in exploitation obtained from the ADAPT assessment model, an exploitation fraction index (EFI) was calculated as the ratio of the catch to the spring survey index of abundance. This index is independent of the assessment model, being based only on data. It is an index because the survey index was used rather than a survey estimate of total abundance.

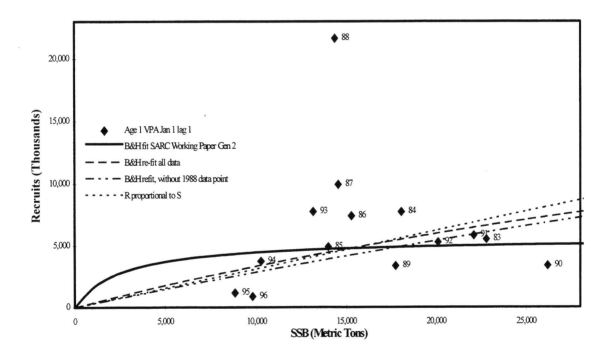

FIGURE 2.1 Gulf of Maine cod spawning stock biomass (SSB) and recruitment from original VPA 1983-1996. SOURCE: NEFSC, 1997a.

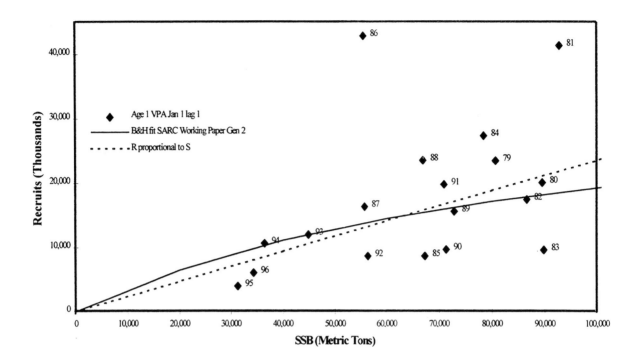

FIGURE 2.2 Georges Bank Cod spawning stock biomass (SSB) and recruitment from original VPA 1979-1996. SOURCE: NEFSC, 1997a.

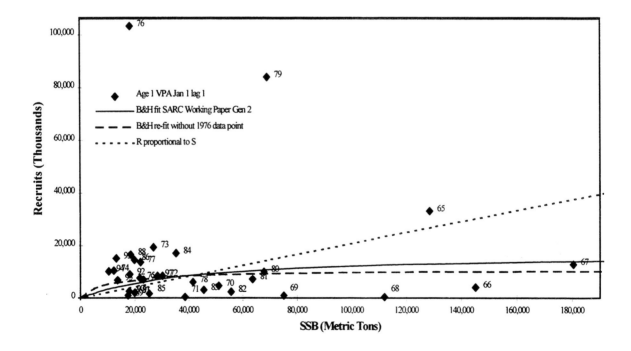

FIGURE 2.3 Georges Bank haddock spawning stock biomass (SSB) and recruitment from original VPA 1965-1996. SOURCE: NEFSC, 1997a.

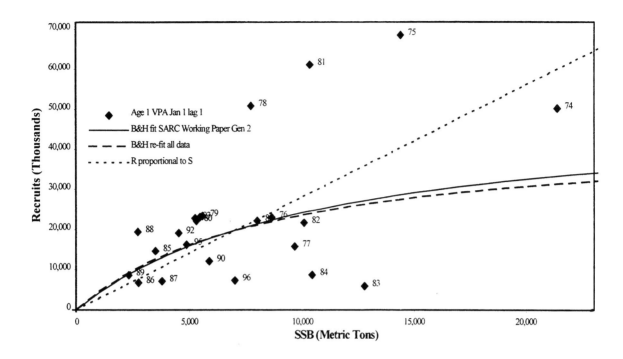

FIGURE 2.4 Georges Bank yellowtail flounder spawning stock biomass (SSB) and recruitment from original VPA 1974-1996. SOURCE: NEFSC, 1997a.

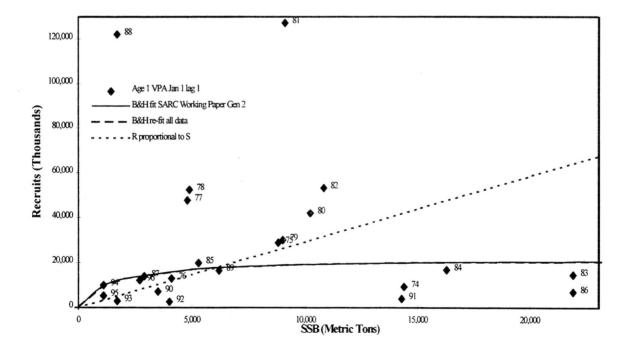

FIGURE 2.5 Southern New England yellowtail flounder spawning stock biomass (SSB) and recruitment from original VPA 1974-1996. SOURCE: NEFSC, 1997a.

Gulf of Maine Cod
(NEFSC, 1997a; pp. 51-107)

Stock Size and Condition

The survey data, ADAPT output, and CPUE data all suggest that the current stock size is well below that of the 1960s and 1970s (see Appendix D; NEFSC, 1997a, 1997b). Standardized landings per unit effort (LPUE) data in 1995-1996 were about 30% of the 1982-1983 values (NEFSC, 1997a, Figure A3). Survey biomass estimates in 1994-1996 were roughly 25% of the 1960s values (NEFSC, 1997a, Figure A4). The spawning biomass estimate for 1996 is 10,700 tons, compared to 24,500 tons in 1982 (Figure 1.2). ADAPT estimates are not available for years before 1982.

Recent Exploitation Rates

All indicators suggest that the exploitation rate remains high. Fishing effort has increased consistently over time; few fish are now found in the catch or surveys in the age 7+ group; and the current catch is moderately high in historical terms (Figure 2.6), whereas indicators suggest that the stock size is low (Figure 1.2). The current F is estimated to be 1.04 (NEFSC, 1997a; see Appendix D).

Have Current Regulations Reduced Fishing Mortality (F)?

The current assessment suggests that F has not been reduced for the Gulf of Maine. The model-independent Exploitation Fraction Index (EFI) in 1995 and 1996 is of the same magnitude as in previous years, suggesting that there has been no major reduction in fishing mortality (Figure 2.7).

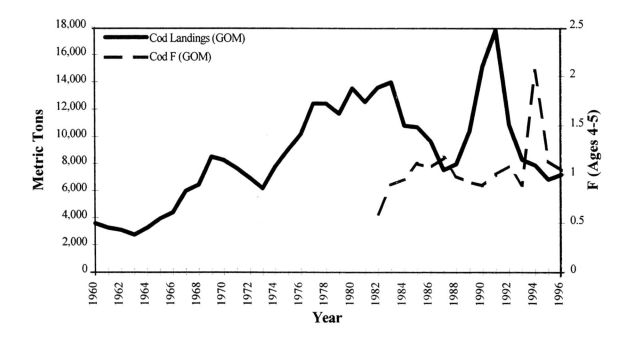

FIGURE 2.6 Commercial landings (metric tons, live) and fishing mortality of Gulf of Maine (GOM) cod (ages 4-5). Based on ADAPT-tuned VPA. SOURCE: NEFSC, 1997a.

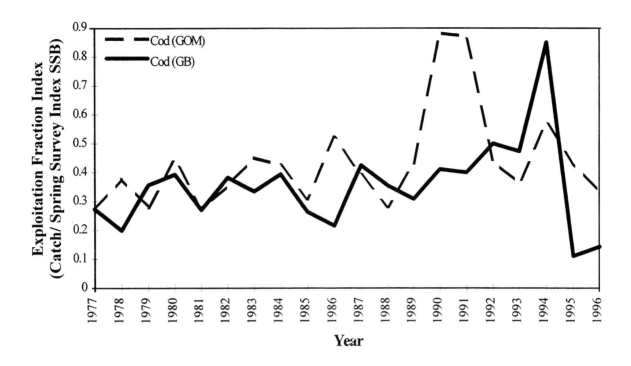

FIGURE 2.7 The ratio of commercial landings to spring survey index spawning stock biomass (SSB) for Gulf of Maine (GOM) and Georges Bank (GB) cod. SOURCE: NEFSC, 1997a.

Will Yield-Per-Recruit Change with Lower Fishing Mortality (F)?

The yield-per-recruit curve is reasonably flat over all target ranges of F. The estimated increase in yield-per-recruit would be perhaps 15% if F were reduced from the current value of 1.04 to F_{max} (F=0.29). There would be associated increases in CPUE and age-range observed in the stock and in the catch. Lower fishing mortality would be likely to decrease the cost of harvesting and would increase spawning biomass per recruit. Spawning biomass per recruit is only about 10% of the unfished level at the current F and would increase to about 25% at F_{max} and 40% at $F_{0.1}$ (which is near $F_{40\%}$).

Will Recruitment Increase with Increasing Spawning Biomass or Decline with Recent High Fishing Mortality (F)s?

The spawner-recruit data for Gulf of Maine cod are inconclusive; therefore, drawing unambiguous conclusions from these data is impossible. NMFS has fit a Beverton-Holt spawner-recruit relationship through the data that suggests relatively constant recruitment over the historical range (1982-1995) of spawning stocks. Estimated recruitments for the last three years are of considerable concern (Figures 2.1, 2.8)—they may reflect a depensatory relationship (lower recruit-per-spawner at lower stock biomass; see the section Alternative Projections for Gulf of Maine Cod below), or they may reflect environmentally driven poor recruitment years. There is no evidence from the stock assessment that rebuilding the spawning stock will result in substantial increases in recruitment over the 1982-1995 levels shown in Figures 2.1, 2.8. However, failing to rebuild the spawning stock may result in a drastic stock collapse. Recent low recruitments may be the first signs of recruitment overfishing (with the caveat that recent recruitment estimates are the most uncertain).

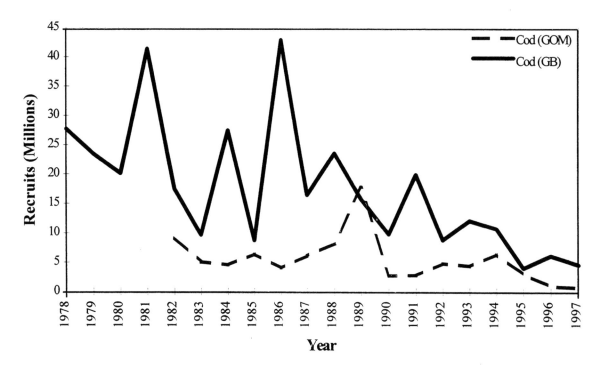

FIGURE 2.8 Recruitment of Gulf of Maine (GOM) cod at age 2 and Georges Bank (GB) cod at age 1 in millions of fish. SOURCE: NEFSC, 1997a.

Alternative Projections for Gulf of Maine Cod

The committee conducted a limited set of simulations to test the effects of various fishing mortalities on stock projections (Figures 2.9-2.13). These simulations were designed to illustrate how stock responses to management measures could be affected under different stock-recruitment scenarios. The committee tested four models of stock-recruitment relationships and the effects of these relationships on stock projections:

1. a Beverton-Holt spawner-recruit model used in NMFS assessments (Figure 2.9);
2. a spawner-recruit model in which recruitment increases in proportion to spawning biomass (Figure 2.10);
3. a spawner-recruit model in which recruitment is constant at the mean historical value, independent of spawning biomass (Figure 2.11); and
4. a depensatory spawner-recruit model in which the ratio R/S increases at low spawning biomass (Figure 2.12). As a result, a stock at low spawning biomass will continue to experience low recruitment on average until spawning biomass increases beyond the depensation threshold.

These figures show that the trend and amount of uncertainty in future projections depends strongly on which spawner-recruit model is used; implications of these results were previously discussed in the section "Evaluating Consequences of Alternative Management Actions" (pp. 31-34). In Figure 2.13, an additional projection is shown for which the constraints on recruitment used in the NMFS analysis are removed (see Appendix F for details). The amount of variability increases by removing the constraints, thereby showing increased uncertainty in future projections.

47

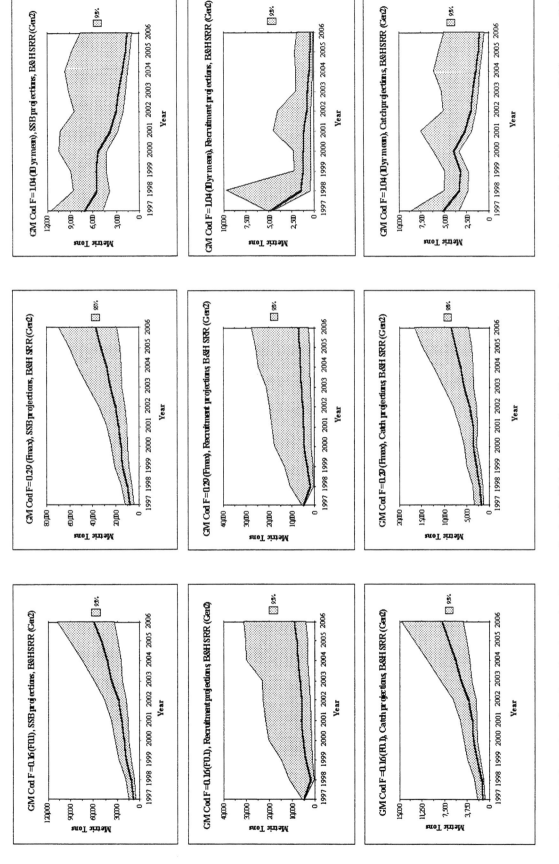

FIGURE 2.9 Results of stochastic projection runs for Gulf of Maine cod using a Beverton-Holt stock-recruitment model and three target fishing mortalities ($F_{0.1} = 0.16$, $F_{max} = 0.29$, and a target fishing mortality $F = 1.04$ equal to the 10-year mean).

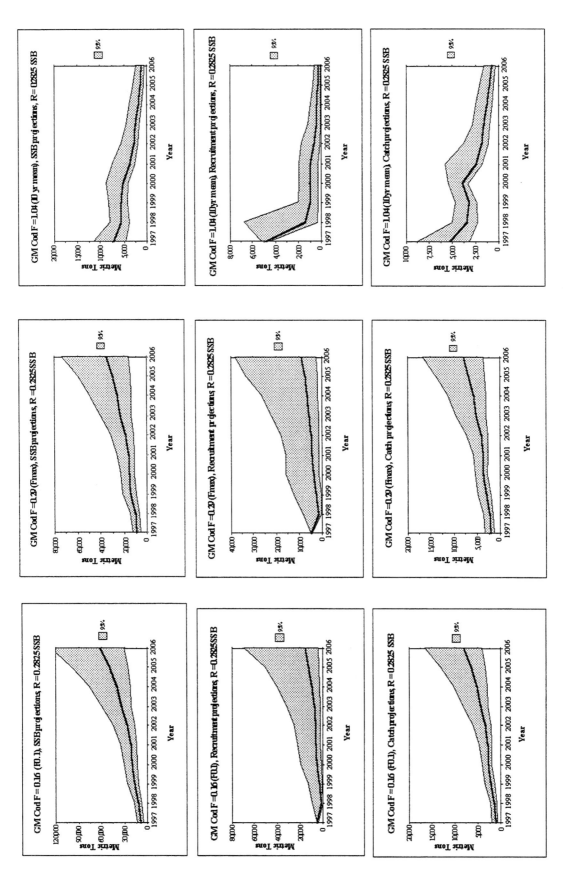

FIGURE 2.10 Results of stochastic projection runs for Gulf of Maine cod using a stock-recruitment model in which R is proportional to S ($R = 0.2825$ SSB) and three target fishing mortalities ($F_{0.1} = 0.16$, $F_{max} = 0.29$, and a target fishing mortality $F = 1.04$ equal to the 10-year mean).

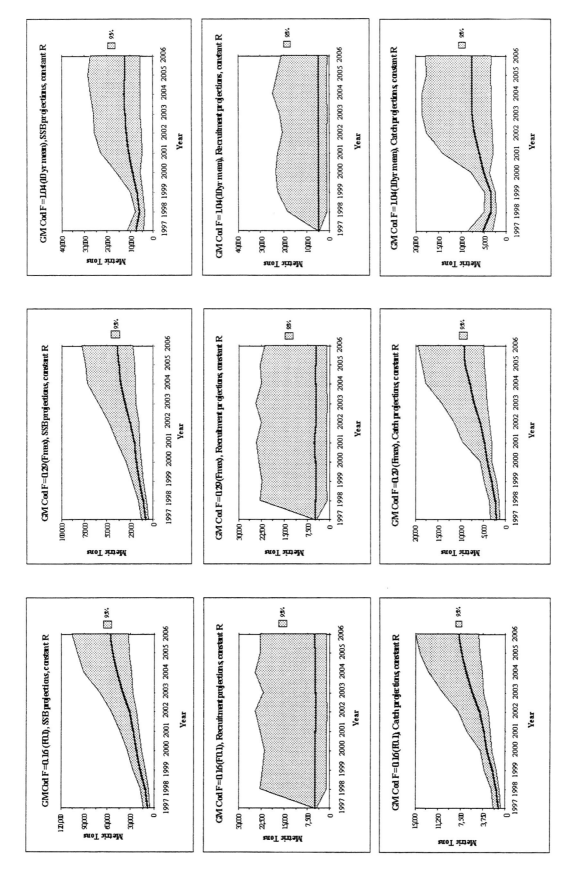

FIGURE 2.11 Results of stochastic projections for Gulf of Maine cod using a constant recruitment stock-recruitment model and three target fishing mortalities ($F_{0.1} = 0.16$, $F_{max} = 0.29$, and a target fishing mortality $F = 1.04$ equal to the 10-year mean).

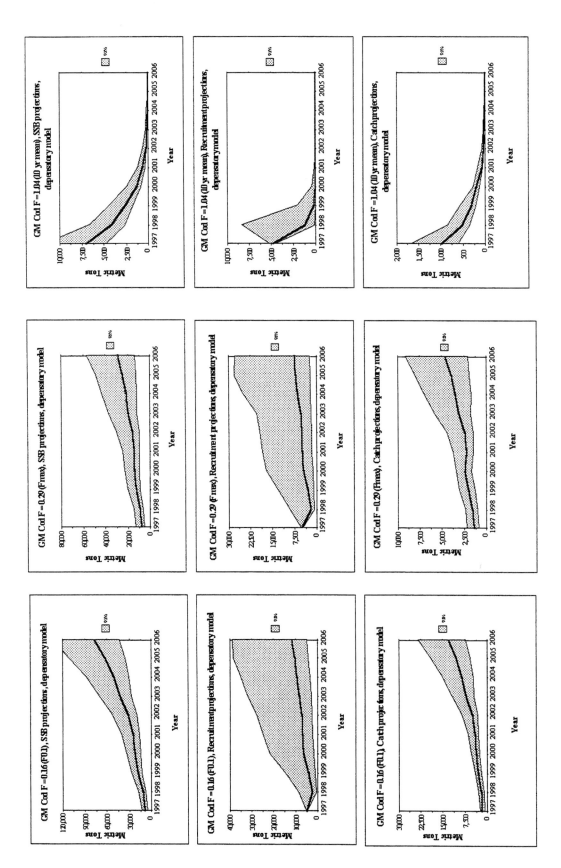

FIGURE 2.12 Results of stochastic projection runs for Gulf of Maine cod using a depensatory S-R relationship recruitment model and three target fishing mortalities ($F_{0.1} = 0.16$, $F_{max} = 0.29$, and a target fishing mortality $F = 1.04$ equal to the 10 year mean).

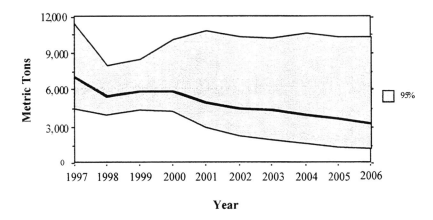

FIGURE 2.13 Results of a stochastic projection for Gulf of Maine cod spawning stock biomass (SSB) using a spawner-recruit model with recruitment proportional to spawning biomass and a high target fishing mortality $F=1.04$ equal to the 10-year mean and removing the constraints on recruitment used in the NMFS analysis.

Georges Bank Cod
(NEFSC, 1997a; pp. 108-170)

Stock Size and Condition

CPUE and survey trends from NMFS show that the spawning stock biomass of Georges Bank cod, in 1996, was roughly one-fifth the level of the late 1970s (NEFSC, 1997a, Figures B2, B4). The ADAPT estimate of spawning stock at its lowest, in 1994, was roughly one-third of the 1980 estimates, the highest in the historical record (Figure 1.2). Analysis of the long-term catch data (see Appendix F) suggests the stock in 1980 amounted to perhaps one-half the potential unfished spawning stock biomass (SSB). This indicates that the 1994 spawning stock biomass was likely in the range of 10-20% of unfished stock size. The latest ADAPT runs indicate some rebuilding of spawning stock (Figure 1.2). Canadian data from the most recent stock assessment in the 5Zj,m areas show a similar trend in the SSB of Georges Bank cod (Hunt and Buzeta, 1997).

Recent Exploitation Rates

The age distribution and effort suggest a high fishing mortality rate until 1995, which is consistent with the ADAPT outputs (Figure 2.14). The current F is estimated to be 0.18 (NEFSC, 1997a), and estimates of F in 1995 and 1996 are much lower than in previous years (Figure 2.14).

Have Current Regulations Reduced Fishing Mortality (F)?

Reductions in effort and landings (Figure 2.14) are consistent with a significant drop in F in the last two years. Again, this drop is also indicated by the ADAPT output. The model-independent EFI shows a strong drop in the last two years (Figure 2.7), suggesting that fishing mortality has been reduced.

The Canadian assessments show a similar decrease in F, corroborating the U.S. assessment results, although there may have been a slight increase in F in the most recent year. Nevertheless, the exploitation rate in the last two years is dramatically lower than the 1978-1994 exploitation rates (Hunt and Buzeta, 1997).

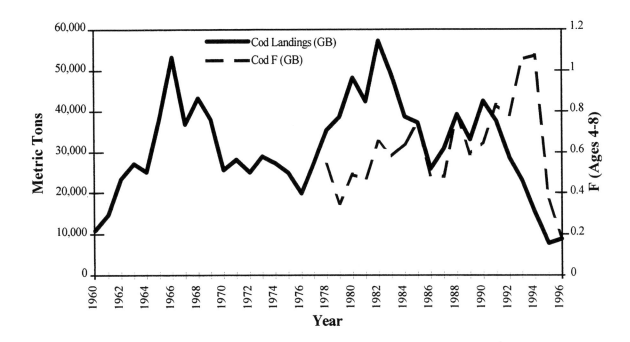

FIGURE 2.14 Commercial landings (metric tons, live) and fishing mortality of Georges Bank (GB) cod (ages 4-8). Based on ADAPT-tuned VPA. SOURCE: Serchuk et al., 1994; NEFSC, 1997a.

Will Yield-Per-Recruit Change with Lower Fishing Mortality?

The yield-per-recruit analysis indicates minor gains in yield-per-recruit by reducing F from high values before 1995 to F_{max}. An expected reduction of less than 10% of the maximum is predicted when F is reduced to $F_{0.1}$ (see Appendix E for definition). Spawning biomass per recruit is only about 15% of the unfished level at the high values of F before 1995 and increases to about 25% at F_{max} and 40% at $F_{0.1}$ (which is near $F_{40\%}$).

Will Recruitment Increase with Increasing Spawning Biomass or Decline with Current Fishing Mortality (F)?

The spawner-recruit analysis for Georges Bank cod suggests a near linear spawner-recruit relationship (Figure 2.2), and as in the case of Gulf of Maine cod, U.S. data show that recruitments in the most recent years are particularly weak (Figure 2.8). Canadian data also show similarly poor recruitment in the most recent years (Hunt and Buzeta, 1997). This analysis suggests the possibility of depensation at low spawning stock sizes; it also provides support for the hypothesis that larger spawning stocks will result in larger recruitments. The committee suggests that increases in recruitment are not likely at spawning stock levels higher than those seen in the early 1980s. The possibility of depensation raises the concern of stock collapse if spawning biomasses were to decline below the levels of 1994.

Georges Bank Haddock
(NEFSC, 1997a; pp. 171-223)

Stock Size and Condition

The haddock stock is much less abundant now than it was before 1960. Although survey and CPUE data do not extend back before 1960, catch and survey data from the early 1960s provide evidence of higher historical levels of abundance (Figure 1.2, Appendix F; NEFSC, 1997a, Figures C2, C5). According to a long-term VPA (1931-1986), the spawning stock was between 100,000 and 300,000 metric tons prior to 1960 and was as low as 11,000 metric tons in 1993. The most recent NMFS assessment shows an increase from this historical low in abundance to 32,400 metric tons in 1996 (NEFSC, 1997a). Canadian results from the smaller 5Zj,m area assessment show a similar increase in spawning biomass since 1993 (Gavaris and VanEeckhaute, 1997). It appears that the recent small increases in spawning biomass (Figure 1.2) are due to lower fishing mortality on the existing biomass; recent recruitments for this stock are low (Figures 2.3, 2.17). The committee concludes that the Georges Bank haddock stock has collapsed.

Recent Exploitation Rates

The age structure and total effort suggest that F was very high prior to 1995. Current regulations have reduced F in 1995 and 1996 (Figure 2.15). The current F is estimated to be 0.18.

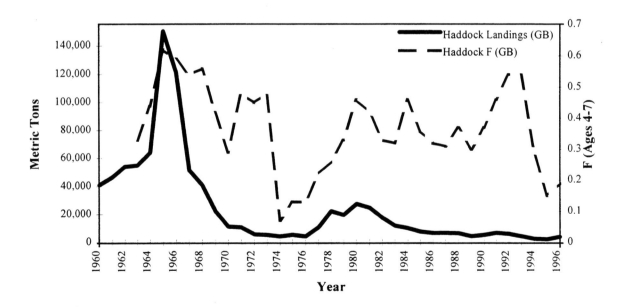

FIGURE 2.15 Commercial landings (metric tons, live) and fishing mortality of Georges Bank (GB) haddock (ages 4-7). Based on ADAPT-tuned VPA. SOURCE: O'Brien and Brown, 1997; NEFSC, 1997a.

Have Current Regulations Reduced Fishing Mortality (F)?

Reductions in effort and implementation of closed areas are consistent with the drastic declines in fishing mortality that emerge from ADAPT runs. The model-independent EFI shows a strong drop in the last two years (Figure 2.16), supporting a decrease in fishing mortality.

Will Yield-Per-Recruit Improve with Lower Fishing Mortality (F)?

The yield-per-recruit analysis shows that the expected yield continues to increase slightly when fishing mortality increases to more than one. An expected reduction of about 20% of the maximum yield is predicted when F is reduced to $F_{0.1}$. Spawning biomass per recruit is less than 10% of the unfished level at the highest values of F shown on the yield-per-recruit graph. Spawning biomass per recruit increases to about 40% at $F_{0.1}$ (which is near $F_{40\%}$).

Will Recruitment Increase with Increasing Spawning Biomass or Decline with Recent High Fishing Mortality (F)?

The spawner-recruit analysis for haddock is particularly complex. The data show that recruitment has varied since 1968 (Figures 2.3, 2.17, Appendix F). It has fluctuated without a trend about an average of 13.5 million recruits, with two large year classes in 1975 and 1978. Thus, at first sight, if the data prior to 1968 are ignored, there are no indications that higher spawning stock biomass produced larger recruitment from 1968 to 1996. However, since 1968, the spawning stock biomass has never been higher than 80,000 tons, whereas it had never been below that value from 1931 to 1967, the period during which substantially higher recruitment was observed. Potentially, there would be significant losses in not rebuilding the stock if 80,000 tons of spawning stock biomass were in fact a real biological threshold below which the average productivity is substantially lower. Of the two strong year classes produced since 1968, the 1975 year class was apparently a result of the particularly good survival of the spawning products from a small spawning stock biomass, whereas the 1978 spawning stock biomass was one of the highest during 1968-1996. As a result of the strong 1975 and 1978 year classes, the SSB remained higher than 40,000 tons from 1977 to 1982, but no other strong year classes were produced during that period. Recent studies (Marshall and Frank, 1994; Chambers and Trippel, 1997) strongly suggest that reproductive success may be a function of the quality of spawners, not just their quantity. Therefore, it is possible to imagine a scenario in which the spawning stock biomass during 1979-1982 consisted mostly of first-time spawners whose spawning products have a low probability of survival.

There are indications that the higher recruitments recorded prior to 1968 may have been produced from two major spawning aggregations, one on the northeast peak of Georges Bank and the other in Nantucket Shoals-West Gulf of Maine (McCracken, 1960; Grosslein, 1961; Clark et al., 1982). One hypothesis is that the Nantucket Shoals/West Gulf of Maine spawning unit may have been severely overexploited and that a larger proportion of the recruits are now produced from the northeast peak spawning unit. Recent work suggests that elimination of local stocks is a major problem for Gulf of Maine cod (Ames, 1997), and that serious attention needs to be given to this situation. Hydroclimatic changes have occurred in this area, but their magnitude has been substantially lower than in the northern areas off Newfoundland and Labrador, where they have been invoked as one of the causative factors in stock collapses (Myers et al., 1996). However, haddock in this area are at the southern limit of their distributional range, and small hydroclimatic changes may have a proportionately greater effect on stock dynamics.

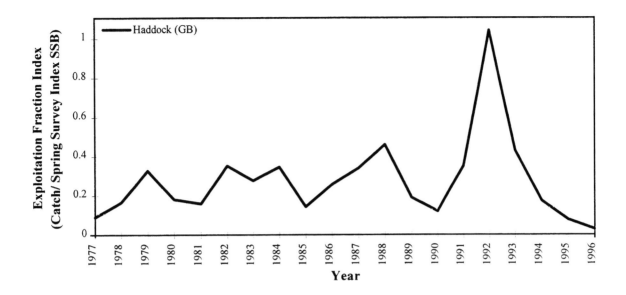

FIGURE 2.16 The ratio of commercial landings to spring survey index spawning stock biomass (SSB) for Georges Bank (GB) haddock. SOURCE: NEFSC, 1997a.

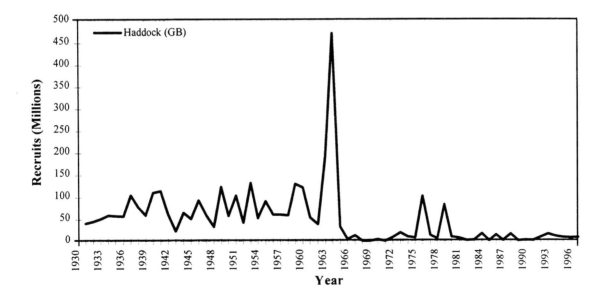

FIGURE 2.17 Recruitment of Georges Bank (GB) haddock in millions of fish at age 1. SOURCE: NEFSC, 1997a.

Although it is possible that the low recruitments since 1968 have been caused primarily by hydroclimatic or environmental changes, it would seem to be extremely important to rebuild the spawning stock biomass to more than 80,000 to 100,000 metric tons. It should be noted that the historically smaller haddock stock on Browns Bank and in the Bay of Fundy (Northwest Atlantic Fisheries Organization Division 4X) has consistently produced higher year classes since 1978, perhaps because spawning stock biomass there has been maintained closer to an optimal value.

Georges Bank Yellowtail Flounder
(NEFSC, 1997a; pp. 224-259)

Stock Size and Condition

The four survey indices of stock size available for Georges Bank yellowtail flounder indicate that the lowest stock sizes were observed from 1987 to 1989 (Figure 2.18). The U.S. spring and fall surveys, which have been conducted since the 1960s, suggest that stock sizes during this period were considerably smaller than those of the late 1960s. The trends of the various indices are inconsistent for the recent period: the scallop and Canadian surveys suggest increases in stock biomass since at least 1993 (Neilson et al., 1997), the 1995 U.S. spring and fall surveys indicate that there has not been any increase in stock size in recent years. VPA results (NEFSC, 1997a, Figure D14) also suggest substantially lower biomass in the late 1980s (less than 3,000 tons in 1987-1988) than in the early 1970s (21,000 tons in 1973).

Recent Exploitation Rates

The high fishing mortality rates estimated by the VPA prior to 1995 are consistent with the almost total lack of individuals older than 5 years observed in the survey data. The current F is estimated to be 0.25, and estimates in 1995 and 1996 are much lower than in previous years (Figure 2.19).

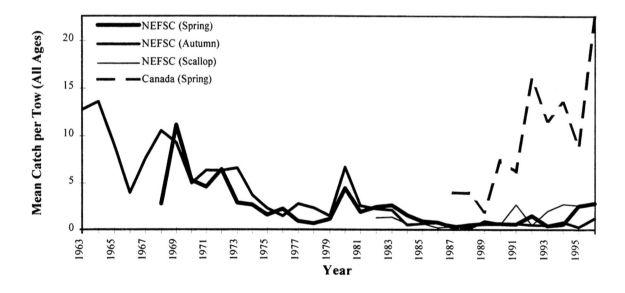

FIGURE 2.18 Commercial survey indices for Georges Bank yellowtail flounder (mean catch [kg] per tow for all age classes).

General Review of Northeast Groundfish Stock Assessments 57

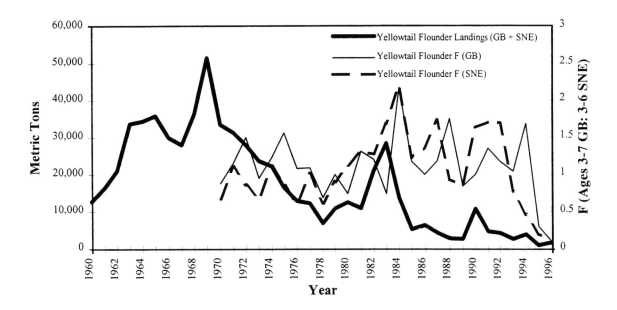

FIGURE 2.19 Commercial landings (metric tons, live) and fishing mortality of Georges Bank (GB: ages 3-7) and southern New England (SNE: ages 3-6) yellowtail flounder. Based on ADAPT-tuned VPA. SOURCE: yellowtail flounder (GB): NEFSC, 1994b, 1997a; yellowtail flounder (SNE): NEFSC, 1994c; 1997a.

Have Current Regulations Reduced Fishing Mortality (F)?

Reductions in effort and implementation of closed areas are consistent with the drastic declines in fishing mortality that emerge from ADAPT runs. The model-independent EFI shows a strong decline in the last two years (Figure 2.20), suggesting that fishing mortality has been reduced. However, these data should be interpreted with care: there are indications, at least for the Canadian survey, that the catchability or availability may have gone up in 1996.

Will Yield-Per-Recruit Improve with Lower Fishing Mortality (F)?

The yield-per-recruit graph is shown in the SARC advisory report (NEFSC, 1997b). Cadrin et al. (1997) showed F_{max} = 0.6 and a 12 % reduction in predicted yield per recruit when F is reduced to $F_{0.1}$ = 0.24. The reduction may be inconsequential if recruitment increases due to lower F (see next section). Spawning biomass per recruit is only about 10% of the unfished level at the high values of F before 1995 and increases to about 20% at F_{max} and 40% at $F_{0.1}$ (which is near $F_{40\%}$).

Will Recruitment Increase with Increasing Spawning Biomass or Decline with Current Fishing Mortality (F)?

On average, larger spawning stock sizes have produced more recruits in Georges Bank yellowtail flounder, and both Canadian and U.S. data indicate that increasing spawning stock sizes should provide a higher probability of producing good year classes (Neilson et al., 1997). All four strong year classes

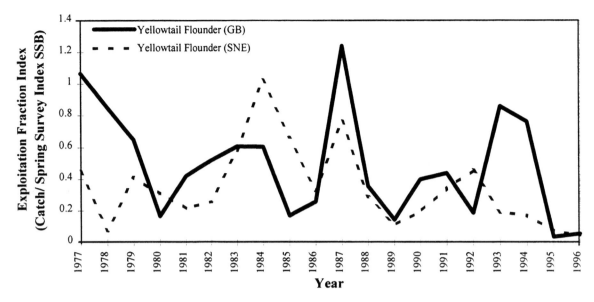

FIGURE 2.20 Ratio of commercial landings to spring survey index spawning stock biomass (SSB) for Georges Bank (GB) and southern New England (SNE) yellowtail flounder. SOURCE: NEFSC, 1997a.

since 1973 have been formed from spawning stocks in excess of 7,000 metric tons (Figures 2.4, 2.21, Appendix F). The 1995 year class is among the weakest in the series, and it was produced from a spawning stock close to 7,000 metric tons. If fishing mortality had not been reduced, spawning biomass would have been lower, and there is no indication whether this would have resulted in even poorer recruitment. Thus, the major choice is between a future similar to the recent past or larger recruitments based on improved spawning stocks. It would be particularly important, in this case, to extend the assessment periods back to the early 1960s, when survey catch rates indicate substantially higher adult biomass than estimated for 1973-1996, the period considered in the current assessment (see Appendix F).

Southern New England Yellowtail Flounder
(NEFSC, 1997a; pp. 260-290)

Stock Size and Condition

Catch and survey data and ADAPT results all show a major decline in the abundance of southern New England yellowtail flounder (Figures 1.1, 1.2; NEFSC, 1997a, Tables E1, E8). Autumn survey abundances for the mid-1990s are less than 5% of the values observed in the late 1960s (Appendix F). Catch data show similar strong reductions (see Figure 1.1), and recruitment has been weak for a number of years (Figure 2.21). The committee considers the southern New England yellowtail flounder stock to have collapsed.

Mid-1990s Exploitation Rates

A truncated age distribution indicates high exploitation rates in the 1990s consistent with model outputs. The current F is estimated to be 0.12 (Figure 2.19).

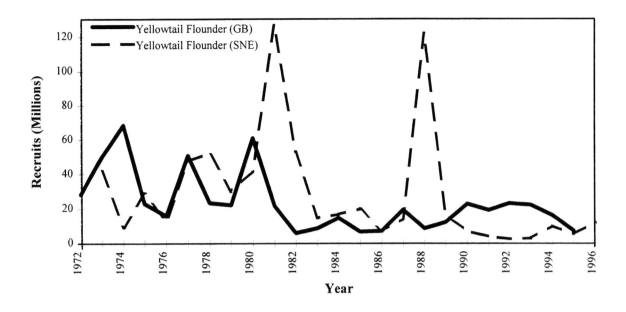

FIGURE 2.21 Recruitment of Georges Bank (GB) and southern New England (SNE) yellowtail flounder in millions of fish at age 1. SOURCE: NEFSC, 1997a.

Have Current Regulations Reduced Fishing Mortality (F)?

Reductions in effort and implementation of closed areas are consistent with major declines in fishing mortality that emerge from ADAPT runs. The model-independent EFI shows a strong drop in the last two years (Figure 2.20).

Will Yield-Per-Recruit Improve with Lower Fishing Mortality (F)?

Yield-per-recruit analysis shows that expected yield continues to increase slightly when fishing mortality increases well beyond 1. An expected reduction of about 15% of the maximum is predicted when F is reduced to $F_{0.1}$. The reduction may be inconsequential if recruitment increases due to lower F (see next section). Spawning biomass per recruit is about 20% of the unfished level at the highest values of F shown on the yield per recruit graph (which are near the levels of F before 1995). Spawning biomass per recruit increases to about 40% at $F_{0.1}$ (which is near $F_{40\%}$).

Will Recruitment Increase with Increasing Spawning Biomass or Decline with Current Fishing Mortality (F)'s?

Spawner-recruit data show that recruitment has been fluctuating without a clear trend over a broad range of spawning stocks, with indications that the most recent years (at low spawning stock biomass) have produced poor recruitments (NEFSC, 1997a; Overholtz et al., 1997; see Appendix F). One of the largest year classes, 1987, was formed from a small spawning stock. Most of the largest year classes, 1976-1981, came in a sequence of years. Recruitments larger than the recent average may not occur from increased spawning stock sizes. Except for the 1987 year class, no strong year class has been produced at spawning stock biomass less than 5,000 metric tons. So there are indications that spawning stock biomass should be kept greater than 5,000 metric tons. In addition, most of the largest year classes (1976-1981) came in a sequence of years. Thus, if only the period covered by the assessment is examined, the two most likely recruitment hypotheses are: (1) although strong recruitment is not necessarily associated with the largest spawning stocks but rather with favorable environmental conditions, recruitment will decline if high F is maintained; and (2) recruitment would stay reasonably unchanged if F is maintained. However, earlier survey catch rates at age that extend back to 1963 (NEFSC, 1997a; see Appendix F) indicate that much larger year classes might have been recruited in the 1960s when biomass was substantially larger. The appearance of large year classes in years when biomass was greater would provide support for a third hypothesis, namely, that stronger recruitments may be possible under favorable environmental conditions when spawning biomass is higher.

3

REVIEW OF DETAILS OF NORTHEAST GROUNDFISH STOCK ASSESSMENTS

Nature puts no question and answers none which we mortals ask. She has long ago taken her resolution.

Henry David Thoreau

This chapter contains detailed technical comments about the stock assessment process. Although incorporating the suggestions contained in this chapter would improve assessments, the committee believes that the current assessments are still valid for making management decisions.

DATA

As mentioned in Chapter 2, the National Marine Fisheries Service (NMFS) has documented the data and an array of information collected over a long period of time that are used in assessments. One notable feature of the Stock Assessment Review Committee (SARC) process is that the evaluation of data sources is explicit and leads to research recommendations for improvements in data collection and documentation (see Appendix D). Some of the recommendations that the committee feels are especially critical are repeated in this report, as well as additional recommendations regarding data collection and treatment.

Dealer and Vessel Data

Problems related to collecting catch and landings information are detailed in the SARC report (NEFSC, 1997a). The new system put into place in 1994 contains a major structural flaw: dealer reports and vessel trip reports cannot be uniquely linked. The system requires a thorough audit and overhaul to fix this and other problems. There needs to be further efforts to verify reported landings. There are several ways to attempt to do this, for example, sociological studies of fishermen's behavior, experimental fishing using gear that the harvesters are using in the area they are fishing, and statistical analyses (Myers et al., 1997).

Misreporting of landings is usually a significant issue only when fisheries are managed by setting a total allowable catch. In the U.S. fishery, there is no total allowable catch set. Most Canadian fisheries have a total allowable catch specified, but their enforcement system may adequately deal with misreporting.

Observer Program

Apparently, information obtained by independent observers aboard fishing vessels is minimal at best; few trips have been observed in recent years. In 1994, 1995, and 1996, a total of 22, 16, and 18 trips, respectively, were observed in the entire otter trawl fleet (DeLong et al., 1997). If fishery regulations continue to restrict harvesters from catching the number of fish they are accustomed to catching, then logbooks may not contain accurate information on levels of catch, location, discards, and so on. Often, the only solution is to station observers on fishing vessels to collect accurate information. Evaluation of the need for a strengthened observer program should be a part of the stock assessment process.

Disaggregated Catch Per Unit Effort

Catch per unit effort (CPUE) has not been used in assessments of the New England groundfish stocks since 1994. With the current quality of logbook data and the various restrictions that recently have been imposed on fisheries, the skepticism about the usefulness of current aggregated catch-and-effort data in constructing CPUE series as expressed by NMFS and the SARC is appropriate. However, harvesters have a greater trust in the data that they themselves provide, and therefore an effort should be made to validate and use CPUE data.

With disaggregated catch-and-effort data, the CPUE series might, however, provide valuable information about the spatial distribution of effort and abundance. Such series might be of value in the regular assessments of stocks and for monitoring purposes. Their value as instruments for monitoring stocks and fisheries depends on data quality, and on the time lag between the collection of new data and the revision of the CPUE index. With an appropriate system for gathering and analyzing catch-and-effort data, this time lag could be shortened, thus helping to improve fisheries monitoring.

To obtain valid CPUE series, changes in fishing technology, fishing competence, and restrictions on effort must be accounted for in the analysis. One approach is to disaggregate the data not only by vessel, but also by skipper and management events. The idea is to focus on periods with constant technology (e.g., same gear, same engine), constant fishing competence (e.g., same skipper and key crew), and same external conditions (e.g., management regime with respect to closed areas and periods, days at sea limitation, rules for discards and bycatch). The catch series will be highly variable within each such period, but by analyzing all spells together, in a generalized linear model, a CPUE series related to relative abundance might be recovered (Hilborn and Walters, 1992).

To obtain data of sufficient quality for disaggregated CPUE analysis, a subset of fishing vessels could be delegated to provide more detailed logbook data than are recorded in the mandatory logbooks. With reliable and detailed catch by time and area, such disaggregated fishery-based data might also be of value for types of analysis other than CPUE series.

Handling Zeros When Estimating Door Conversion Factors and Other Data

To weight data from surveys with different trawl doors (BMV and Polyvalent) equally, a conversion factor has to be estimated. Because zero catch occurs with substantial frequency during the surveys, an *ad hoc* approach was taken in the NMFS analysis. In experimental data, pairs of hauls with trawls using the two types of doors, and pairs with zero catch for either of the two trawls were excluded.

The handling of zeros is also a problem in other analyses. Using a root normal rather than a lognormal distribution might be a simple fix, because the square root transformation works with zero values. In addition, these data could be treated in a statistical model along the lines of Coe and Stern (1982).

Survey Coverage

A major factor in deciding the accuracy of stock assessments is the variability in survey indices. When considering total abundance, the survey variability is directly related to the number of stations considered. The present surveys are somewhat lacking in coverage (Azarovitz, 1981), and considerable benefits potentially could be obtained by increasing survey coverage. For example, in the 1996 spring survey there was a single tow with dramatically higher catches of haddock (Figure 3.1). Increasing the number of stations in the surveys is of major importance. Reallocating stations will not achieve the same effect. Increasing the total number of stations sampled, in conjunction with improved reallocation of the number of samples in each stratum, will increase the precision of the mean and decrease the standard error (Gunderson, 1993). In particular, a portion of the new stations should be allocated to the closed areas, where higher fish densities should be expected if the closed areas are effective. The current number and spatial allocation of fishing sets in the closed areas will not allow an assessment of the effects of closed areas with any confidence. This lack of assessment becomes quite critical when the utility of closed areas in reducing fishing mortality is to be evaluated.

Even though reallocation of stations may not lead to improvements, reevaluation of the survey design could lead to improvement. The use of adaptive sampling (Thompson and Seber, 1996) could lead to a significant improvement in survey design efficiency. Therefore it would be fortuitous for NMFS to examine alternative survey designs which may improve the precision of its surveys.

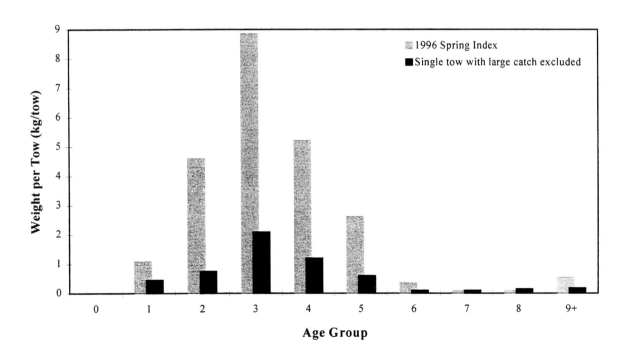

FIGURE 3.1 Mean number of haddock per tow at age captured in the Georges Bank strata sets (offshore strata 01130-1250, 01290-01300) during spring 1996 research vessel survey conducted by the Northeast Fishery Science Center. SOURCE: Adapted from NEFSC, 1997a.

Age Sampling

Age readings are commonly a source of serious problems in stock assessments. The samples taken for age reading fall into two categories: age samples from the commercial fishery and age samples from the surveys. Typically, catches in numbers at age form an important basis for stock assessments, and it is important that the age samples provide accurate information on the age composition in the catches.

Because age samples are commonly taken as a fixed proportion of the tons landed, it is clear that when catches are low, it will be very difficult to keep track of the relative size of year classes if the sample sizes are allowed to become very small as they were for some groundfish species (Table 3.1; NEFSC, 1997a; Brown, 1997; Cadrin et al., 1997; Mayo, 1997; O'Brien, 1997; Overholtz et al., 1997).

For example, a sample size of only 200 fish for a species with two important age groups will result in fairly inaccurate estimates of the proportion in each one. An optimistic estimate of the standard deviation in an important age group could be obtained using the binomial distribution, whereas more appropriate simulation procedures can also be used. To keep track of the age composition, follow year classes, and avoid outliers in fishing mortalities based on virtual population analysis (VPA), it may be necessary to increase sample sizes considerably. A part of future stock assessments should be an evaluation of sample size requirements for ageing to meet specified precision and accuracy goals in stock assessment outputs such as biomass and fishing mortality estimates.

Recreational Sampling

The primary source of marine recreational data for the New England region and the nation is the Marine Recreational Fisheries Statistics Survey (MRFSS) conducted by NMFS with the cooperation of the coastal states. MRFSS is a design-based survey that produces estimates of total effort and catch in directed recreational fisheries over broad coastal areas. It has two components, one to obtain effort data and the other to obtain catch rates. Effort data are obtained from interviewing anglers in households that have been contacted by telephone through random-digit dialing of coastal county phone prefixes. Random-digit dialing is an inexpensive, though inefficient, method to obtain data and results in estimates that can be precise because of the large sample size. Because bias is incurred when anglers self-report their catch rates, these data are combined with data obtained from on-site interviews where the survey clerk sees and measures the catch. Although data quality from on-site surveys is high, they require trained clerks, involve transportation, and result in fewer interviews. Thus, on-site interviews are more costly to obtain, fewer interviews occur, and catch rates are estimated with less precision.

TABLE 3.1 Number of Age Samples Collected for Five Groundfish Stocks 1993-1996

Year	Gulf of Maine Cod	Georges Bank Cod	Georges Bank Haddock	Georges Bank Yellowtail	Southern New England Yellowtail
1993	447	3744	649	N/A	262
1994	665	2332	1259	487	195
1995	662	1326	1234	257	171
1996	1483	1959	1636	395	226

NOTE: Data do not include samples collected from recreational fisheries. SOURCES: Gulf of Maine cod: Mayo, 1997; Table 5; Georges Bank cod: NEFSC, 1997a; Table B6; Georges Bank haddock: Brown, 1997; Table 9; Georges Bank yellowtail flounder: Cadrin et al., 1997; Table 2; southern New England yellowtail flounder: Overholtz et al., 1997; Table 2.

Recreational catch is difficult to measure precisely without incurring considerable cost. In MRFSS, catch is calculated by multiplying catch rates from the on-site surveys and effort obtained through telephone interviews. Catch rates are highly variable because the skill levels of anglers and motivations for catching fish differ greatly. Most anglers catch few fish in a given trip, whereas a few anglers catch many. Thus, when the catch rate per trip is plotted, its distribution is highly skewed. Although an increase in interviews results in better precision, skewness in the confidence intervals is slow to disappear. Skewness is undesirable and difficult to eliminate but does tend to result in conservative management decisions.

Stock assessments are most reliable when catch data are precise. The difficulty in integrating recreational catch data into stock assessment rests largely in the uncertainty of the catch estimates. When recreational catch is a minor part of total catch, this imprecision adds little to the overall uncertainty. When recreational catch is a major or dominant proportion of total catch, uncertainty in the stock assessment predictions is driven by the imprecision of the recreational catch. In the few instances when this occurs, the expense of increasing the number of on-site interviews or developing a specifically targeted survey may be justified.

Only two of the New England groundfish stocks have appreciable recreational catches: Gulf of Maine Georges Bank cod. Recreational catches are incorporated into these assessments but are sufficiently precise to be useful. As the stocks rebuild, the commercial catch will increase with a concomitantly diminished contribution from the recreational catch. Because of changes in the demography of coastal populations, a significant increase in marine angling is not anticipated. Hence, the catch from commercial fishing should increase more than that from angling. Although it is unlikely, if the recreational catch were to increase in these fisheries, the committee would recommend an increase in the number of on-site interviews through increased participation of the New England coastal states.

Recreational fisheries are sampled for biological characteristics of the catch such as species composition, length, and age. Length can be converted to age by using age-length key tables generated by compiling samples of fish lengths and ages. Fish can also be aged by collecting hard parts such as otoliths and scales. This procedure is problematic because it is time-consuming and disfigures the fish in a manner that many anglers find unacceptable. Time is often better spent obtaining another interview than obtaining hard parts. As long as sufficient length samples are taken and the appropriate methodology is used, estimated catch at age will be precise enough to use in stock assessments.

Environmental/Ecosystem Data

Stock assessments would be much improved by the addition of a section related to environmental and ecosystem considerations. This section would describe data series and information related to oceanography and environmental conditions, and perhaps the extent to which these variables explain changes in recruitment of fish populations. For example, in the assessment of Georges Bank haddock, environmental changes or stock depensation (or both) could explain the reductions in recruitment in recent years. The current assessment does not describe these two possibilities.

In addition, information on stomach contents could be included as a means of examining species interactions. Information about the man-induced impacts (e.g., pollution, drilling, and fishing gear) on habitat and environment could be described. Finally, complex interactions among the fish populations, the environment, and fishery management could be explored. The exploration of these interactions is especially valuable because harvesters who spend time on the water believe that these factors are important.

ASSESSMENT MODELS

The following material describes current and alternative approaches to fish stock assessment models. General principles are given first, followed by an outline of the approaches used in current Northeast groundfish assessments and alternative approaches that should be considered. Several of these concerns have been addressed in documents provided to the committee, but others have not. As with other sections of this report, many of the concerns raised here may not change the overall picture of stock status obtained from assessments. They are suggestions to be considered in the future. Some of these suggestions will lead to somewhat different projections and possibly to higher estimates of variance. The use of alternative methods and data sources will strengthen the advice based on these assessments. More elaborate consideration of stock assessment models and their role in fishery management is contained in the National Research Council report *Improving Fish Stock Assessments* (NRC, 1998).

Initial Evaluation of Input Data

Before a formal stock assessment is undertaken, the available data must be studied closely to determine if they are likely to satisfy the general requirements for such data sets. In particular, some estimates of the relative variability and internal consistency of different data sets are necessary to verify the likely accuracy of final population estimates and the relative weight to be given to different data sources.

Simple models, or smoothers, can be used for this purpose. For example, such methods will yield the inherent variability in survey indices at age before any stock assessment is undertaken. This information is important, if only to show which age groups are unreliable and which sets of indices may not be adequate as parts of the assessment. Such an analysis will provide some information on appropriate weighting to be used in any assessments.

Approaches to Assessment Modeling

As with all biological models, several approaches can be taken to fish stock assessment modeling. Most of the models in use around the world are based on the same foundation, but some aspects of implementation vary considerably. In particular, some models are based on the assumption that catches are measured without error and that the age disaggregation of these catches is also without error.

To ensure that the advice based on these assessments is as good as possible, several different models should be used during the assessment process. At the beginning, when an overall picture is being formed, creativity in modeling and flexibility in analysis are of overriding importance. Particularly useful are parallel assessments conducted by different individuals, possibly at different institutions, in order to ensure that the widest possible range of alternatives is considered. Alternative assessments are already considered to some extent, as described below, but further investigations should be encouraged.

Current Model Assumptions

The primary assessment models chosen for the five stocks under consideration are all based on the assumption that catch-at-age data are without error and that survey indices follow a lognormal distribution. It is further assumed that the survey data have the same variance at age on a logarithmic scale for all age groups. Even within the assessment procedure used, it is possible to evaluate the adequacy of most of these assumptions and the effect of changing them.

Alternative Model Assumptions

The present ADAPT models have been tested extensively against other models used around the North Atlantic region and found to perform similarly (NRC, 1998). In other areas, completely different classes of models have been developed with different emphases. These models should be tested to verify the importance of different assumptions. The verification of model assumptions will become particularly important in future years if the current low fishing mortalities continue to prevail.

Alternative assessment methods should be used to consider the effects of including variability of catch data (through a formal statistical model), other sources of stochasticity or uncertainty in the system, and internal estimation of parameters of stock and recruitment models.

Formal statistical models (ADAPT or other) allow statistical evaluation of various concerns. For example, formal statistical tests can be used to evaluate the effects of outliers or even to determine how much they have to be downweighted. Although all formal statistical tests have to be viewed with some caution in these highly nonlinear systems based on data with unknown statistical properties, formal measures are useful when evaluating such effects as outliers.

Similarly, different weightings and transformations should be compared, in order to understand which components of the objective function dominate the estimate of fishing mortality. This should be done by investigating different weighting of the various fleets in each assessment. It is important to know which indices are driving the assessments. The fact that they are now given equal weight does not mean that each has the same influence on the results. If estimates of the variances in each age group are available, these should be used to give initial weights related to the inverse of the variance.

Correlations between age-disaggregated indices of abundance within each year may be a source of concern when estimating stock sizes. The importance of this effect has to be investigated on a case-by-case basis using methods that incorporate the covariance between age group indices from surveys. These correlations can be important in some cases, even yielding much lower estimates of stock size than previously obtained (Myers and Cadigan, 1995). Some Canadian assessments take these correlations into account.

Further Extensions in Time

Several of the assessments use rather short time series. Although this may be adequate for estimating the current stock size, a much longer view of stock sizes through increased use of historical data is needed. For Georges Bank haddock, a VPA back to 1931 provides this perspective (NEFSC, 1997a). Such a historical perspective is particularly needed for Gulf of Maine cod. All of the stock estimates can be extended backward in time through the use of the survey time series. A simple method is to assume the same catchabilities at age for the surveys over time and then to use simple scaling, as shown in Appendix F. However, this method results in much more variable estimates than ADAPT, although it does show a tendency toward higher abundances in the past. A more advanced approach would involve an age-structured assessment method such as Stock Synthesis or Automatic Differentiation Model Builder, which can use total catch and age-specific survey indices for the entire period, and incorporate catch-at-age information for the time period for which it is available. Alternatively, it may be possible to extend the catch-at-age data back in time and continue to use ADAPT.

Natural Mortality

Only a few assessment models allow for the estimation of natural mortality (M) internally. Usually M is assumed to be known from other sources. In recent years, increased emphasis has been placed on models that explicitly incorporate the estimation of natural mortality (e.g., McAllister and

Ianelli, 1997). As a rule, these models are Bayesian, incorporating a prior distribution on many parameters, including the posterior distribution of natural mortality in the output. Taking into account the uncertainty in natural mortality is particularly important with regard to estimations of uncertainty and predictions of stock size.

Model Verification

Having obtained a stock assessment, available procedures should be used to compare the assessment to all possible data sources, in particular, any data sources not used in the assessment. Specifically, because no effort data after 1994 are used in present assessments, to compare the actual number of days at sea or other effort restrictions on the fishing mortality being inflicted on various stocks would be useful.

At the same time, the results of the stock assessment should be compared with qualitative information available from the fleets. Thus, evidence of a sharp reduction in fishing mortality should be considered in conjunction with corroborative evidence from harvesters about whether they really notice the effect of effort limitations. This evidence is needed to prevent cases in which effort limitations have no effect due to either miscalculation of historical effort or evasions of the system.

As discussed in Chapter 2, available indications are that the present management system has indeed reduced fishing mortality for four out of the five stocks as observed in the present assessments. However, alternative analyses are needed to validate the estimates of fishing effort as a routine part of the assessments.

4

SCIENCE AND MANAGEMENT

If fisheries science is to be successful we must learn from and avoid the mistakes of the past. We must recognize that stock assessment involves understanding and making predictions about the response of fishery systems to alternative management actions. We must help managers make "choices" about "dynamic" fishery systems in the face of "uncertainty."

Quantitative Fisheries Stock Assessment: Choice, Dynamics, and Uncertainty
R. Hilborn and C. Walters 1992

The matter in hand is not simply a scientific one. It is a complicated matter, and has its economic as well as its technical side.

An Enemy of the People
Henrik Ibsen

ROLE OF STOCK ASSESSMENT IN THE FISHERY MANAGEMENT PROCESS

Stock assessment is only one component of the fishery management process. However, it is the primary scientific basis for management of the fishery. When reviewing any fishery management process, it is important to clarify which issues are in question. In some cases, stock assessments are challenged when the contentious issue is some other aspect of the management process. Recent research has demonstrated that the uncertainties inevitably associated with stock assessments create opportunities for strategic, self-interested behavior on the part of council members, the fishing industry, politicians, and interested parties (Healy and Hennessey, 1998a, 1998b). This seems to be the case with the Northeast fishery.

The committee met with interested parties at a public hearing in Gloucester, Massachusetts, on July 11, 1997 (see the section "Public Hearing" later in this chapter). It appeared to the committee that the main reasons for questioning the assessments were the strong management measures proposed as a result of the 1994 assessments, not the assessments themselves. Those at the hearing generally agreed with the results of the current assessments, which indicate substantially lower fishing mortality in 1995 and 1996, and increasing spawning stock biomass for four of the five stocks reviewed. For Gulf of Maine cod, where fishing mortality remains high and stock size low, interested parties at the public hearing did not appear to disagree with the current assessment, but some harvesters were concerned that new management measures might further restrict their fishing operations.

Given that the inherent uncertainty of stock assessments leads to concerns about their central role in the fishing management process, there are two conditions that must be met for stock assessments to be useful in management. First, managers must trust that the uncertainty is within acceptable bounds. Second, managers must act on this trust and use the information in a timely manner.

Uncertainty in Stock Assessments

Uncertainly in stock assessments can be introduced by a variety of factors, including limited survey information, lack of catch and discard data, and limited or incorrect size and age information, as described in Chapters 2 and 3. One method of describing the uncertainty is by showing the coefficients of variation for survivorship estimates. The coefficient of variation (CV) defines the level of variation in an estimated mean value: high CV indicates a high degree in variation; low CV indicates a low degree of variation. Coefficients of variation of the survivor estimates by age for each of the five stocks assessed are given in Table 4.1.

These CVs are typical of age-based assessments calibrated with two to three survey-based stock size indices. As indicated in "Evaluating the Consequences of Alternative Management Actions" (Chapter 2), the uncertainties in stock projections can even be greater than suggested by the CVs in Table 4.1. The fishery management tools implemented for the Northeast groundfisheries do not now involve the setting of total allowable catches (TACs). Uncertainties of this magnitude might jeopardize the credibility of a TAC-based management system in which TAC is adjusted yearly in response to changes in stock size in an attempt to maintain a given exploitation rate or keep the biomass above a given threshold. The principal management measures used in Amendment 7, days at sea and closed areas, do not demand the amount of precision in assessments that TAC management systems do. However, the committee suspects that effort controls of the sort used here will lead to improvements in efficiency by harvesters, which will eventually negate the intended reductions in fishing effort. Therefore, mechanisms to monitor and assess fishing efficiency will be necessary. Furthermore, it is now not possible to predict what future levels of fishing mortality will be, based on current management regulations. Perhaps such predictions will be possible after the current regulations have been in place for a number of years.

Regardless of the management system, the uncertainty in assessments can be reduced by estimating total removals more reliably; collecting more biological samples to characterize the catches; reducing the variance of survey-based stock size indices; and perhaps, deriving stock size indices from commercial or recreational fisheries. It is clearly impossible to implement a science-based decisionmaking process without collecting and analyzing information. The costs involve normal expenditures that must be borne in any kind of business that relies on analyses for decisionmaking. Data collection and analysis are particularly required when individual enterprises benefit from resources that collectively belong to the nation.

Timely Use of Stock Assessment Information

Stock assessments are irrelevant if the information provided is not used in a timely manner by fishery managers. In the case of Northeast groundfish, strong management measures appear to be justified by the stock assessments performed for a number of years. Estimates of fishing mortality were very high and increasing, estimates of spawning stock biomass were very low and decreasing, and some estimates of recruitment were at or near the lowest on record, indicating that the harvesting strategy was not sustainable. Management actions prior to 1995 had little effect on controlling fishing mortality, so that only the series of actions commencing with Amendments 5, 6, and 7 seem to have been able to reduce fishing mortality to levels that may be sustainable. The 1997 assessments show some positive

TABLE 4.1 Coefficient of Variation (percent) of Survivor Estimates by Age

Age	Gulf of Maine cod[a]	Georges Bank cod[b]	Georges Bank haddock[c]	Georges Bank yellowtail[d]	Southern New England yellowtail[e]
1		52	62		
2	46	33	40	53	71
3	31	27	31	35	44
4	31	27	29	32	36
5	41	28	27	22	34
6	57	30	26		
7		32	33		
8		33	34		
9					

[a] Table 17; Mayo, 1997.
[b] Table 20; O'Brien, 1997.
[c] Page 15 (Appendix); Brown, 1997.
[d] Table 19; Cadrin, et al., 1997.
[e] Page 55; Overholtz, et al., 1997.

signs of increases in spawning biomass and a widening age composition in the population and catch, although these estimates are uncertain and could change as new information becomes available. It is difficult to say which of the regulations (e.g., which closed areas, what reduction in days at sea) have contributed most to the estimated reduction in fishing mortality. Future stock assessments will have to address this question. In assessing the value of timely stock assessment information it is important to note that the failure of managers in Canada to incorporate scientific information appropriately played a central role in management decisions that helped precipitate the collapse of Atlantic cod in eastern Canada (Hutchings et al., 1997).

STRATEGIC THINKING AND PROCESS

In marine wild capture fisheries, such as the ones under review, fisheries management is often more about managing human activities than managing fish, because there are substantial ecological, economic, and technical constraints on marine fish enhancement. In an ideal fisheries management system, the management agency and interested parties specify social, economic, and biological management objectives to be reached. The scientific assessments then help scientists and managers evaluate the probabilities of reaching these management objectives under various scenarios.

In the Northeast fishery, the specification of both long-term and short-term goals is the purview of the New England Fishery Management Council. Biological limit reference points and the definitions of overfishing and biomass thresholds are now used as de facto objectives. This interim strategy is designed to start the rebuilding process for fish stocks, but at some point, long-term management goals will have to be developed. It is quite possible that stringent management measures, such as those contained in Amendments 5, 6, and 7, would not have been necessary, had a long-term strategy been formulated and implemented in the 1980s. It is possible that implementation of alternative management measures also could result in stock rebuilding.

Science has a role to play in assisting management with the process of adopting clear social and economic objectives within biological constraints and developing a management plan designed to reach the objectives. The first role is the customary one of stock assessment science: to investigate the biological productivity of fish populations in a changing world and the effect of harvesting on that productivity. The second role is to identify the socioeconomic changes that might occur due to changes in biological productivity and management measures.

Socioeconomic studies have been conducted in the Northeast (Aguirre International, 1997; Clay and Dolen, 1997; OECD, 1997) and have had an influence in the decisionmaking process. One challenge is how to incorporate socioeconomic considerations formally and explicitly into scientific advice and the decisionmaking process. Such a policymaking framework or process would include, to a much larger extent, the participation and concerns of stakeholders in the fishery. Giving stakeholders more say in formulating and implementing policy might mitigate some of the existing tensions between NMFS and harvesters, as well as transfer some of the responsibility for the consequences of policy to those involved directly with exploiting the fishery.

The stock assessment group could undertake a number of analyses, probably in collaboration with a fisheries economist, that might greatly enhance the New England Fishery Management Council's ability to evaluate policy options and set more stable and socially acceptable management policies. Such policies would take a long-term view, with built-in adaptive components; provide interested parties with prior knowledge of what happens when a run of poor recruitment years reduces stocks to dangerously low levels; and enable them to make contingency plans for such events.

Most of these additional analyses that could be undertaken come under the rubric of "scenario experimentation." To provide some idea of the types of risk analyses that might be useful to managers, for example, consider the issue of setting a policy to rebuild stocks. Additional analyses would provide an evaluation of how projections may be affected by both uncertainties in the true values of fishing mortality each year and the form of the stock-recruitment relationship (as mentioned earlier). Specific questions that might be addressed include the following: What if the real level of fishing during the past season is much higher (or lower) than estimated and current regulations are unchanged? What if actual recruitment potential over the next two to three years is much stronger (or weaker) than anticipated, yet current regulations cause a particular proportion of harvesters to abandon the industry during the next season? What if historically high levels of recruitment never return once the stocks are rebuilt, because of fundamental changes to the ecosystem (e.g., changes in food web structures, pollution, or gear damage to the environment)?

For such additional risk analyses to be worthwhile, greater interaction and information exchange between assessment scientists and policymakers may be needed. Scientists can help policymakers develop long-term strategies for managing the fisheries, rather than merely short-term tactics to get the fisheries through their current difficulties. Similarly, policymakers can help scientists in their assessment activities by articulating clear policies that can then be investigated with careful scientific modeling of both assessment and management processes.

PUBLIC HEARING

The committee devoted one afternoon to a public hearing with harvesters and representatives of harvesters' associations, representatives of conservation and environmental organizations, and social scientists studying harvesters. Formal presentations were made by some participants and informal comments were provided by others. Several opportunities were provided during the afternoon for informal discussions between participants and committee members. Some of the salient points raised at this hearing are as follows:

1. Participants generally agreed that the current regulations designed to reduce fishing mortality are working.

2. Environmental groups stressed the importance of continuing regulatory measures leading to the rebuilding of stocks.

3. A number of harvesters and representatives of their associations, as well as sociologists working with them, stressed their concerns that environmental factors not included in stock assessment models may play an important role in determining stock sizes. Pollution and global climate changes were raised as two of the possible environmental factors.

4. Harvesters and their representatives spent a good portion of the public hearing criticizing fisheries management measures, specifically their development and implementation. Although they reported having ample opportunities to speak at hearings and other meetings, they expressed feelings that no one was really listening to them. They reported difficulties in communicating effectively with National Marine Fisheries Service (NMFS) scientists in the past, although they mentioned an improving climate of interaction recently.

5. Harvesters repeatedly stated that the way in which regulations were implemented, often one on top of another, made planning difficult. The harvesters expressed concern that they alone had to pay for the mistakes of the past—both their own and those of government. In particular, they pointed out that when foreign harvesters were excluded when the exclusive economic zone (EEZ) was introduced in the 1970s, various public plans were put in place to increase the capacity of the Northeast fishing fleet. The plans encouraged recruitment of harvesters and increased investments in the industry, evidently in excess of what the fishery could sustain. "Is it right," they ask, "that the harvesters now should bear the burden of the government's overoptimism then?"

Committee Comments on the Public Hearing

The committee's task is to review the scientific basis for stock assessment techniques, and its considerations are necessarily focused on scientific issues rather than social, economic, and management-associated concerns. Thus, although the major points raised at the public hearing have been reported here, the committee cannot directly address those that do not deal with scientific issues of stock assessment. All of the concerns raised at the public hearing, however, have served the useful purpose of reminding committee members of the impact of stock assessments on the community of harvesters. The committee has endeavored to make its scientific review with these human concerns always in mind. It was heartening to learn that there appears to be a general recognition that fishing mortality is decreasing and that stocks are increasing. No one seems to doubt that the current effort to restore depleted fish populations is succeeding. Likewise, the reports of improved communication between harvesters and NMFS scientists may help to reduce misunderstanding about the way stock assessments are produced and used in fisheries management. The committee learned that harvesters will begin to accompany NMFS personnel on scientific surveys; the committee commends this effort and encourages the continuation and extension of similar efforts. Because the regulation of fishing effort affects the livelihood of harvesters and because regulations flow directly from the stock assessments, it is inevitable that harvesters should be concerned about the stock assessment process. Mutual efforts at communication and education can only help reduce tension between harvesters and management.

5

CONCLUSIONS

Science which cuts its way through the muddy pond of daily life without mingling with it casts its wealth to the right and left, but the puny boatmen do not know how to fish for it.

Alexander Herzen

Although the committee believes that the 1997 Northeast groundfish stock assessments were sufficient and appropriate for providing scientific input to management decisions, the following recommendations are suggested to improve the quality of information and advice.

RECOMMENDED ACTIONS

1. Improve the collection, analysis, and modeling of stock assessment data as detailed in Chapter 3. Such improvements could include evaluations of sample size, design, and data collection in the fishery and the surveys; the use of alternative methods for data analysis; consideration of a wider variety of assessment models; and better treatment of uncertainty in forecasting;

2. Improve relationships and collaborations between NMFS and harvesters by providing, for example, an opportunity to involve harvesters in the stock assessment process and using harvesters to collect and assess disaggregated catch per unit effort data;

3. Continue to educate stock assessment scientists through short-term exchanges among NMFS centers so that each center can keep abreast of the latest improvements in stock assessment technologies being used at other NMFS fishery science centers, and at other organizations in the United States or elsewhere;

4. Ensure that a greater number of independent scientists from academia and elsewhere participate in the Stock Assessment Review Committee (SARC) process; where necessary, pay competitive rates for such outside participation to ensure that a sufficient number of the best people are involved in the review;

5. Increase the frequency of stock assessments. As the New England Fishery Management Council intensifies its management of the Northeast fishery, stock assessments may have to be performed more frequently than every three years (the current timing);

6. Consider a wider range of scenarios (e.g., recruitment, individual growth, survival, sub-stock structure, ecosystem, data quality, compliance with regulations, long-term industry response) in evaluating management strategies;

7. Investigate the effects of specific management actions, such as closed area and days at sea limitations, on fishing mortalities and related parameters;

8. Work toward a comprehensive management model that links stock assessments with ecological, social and economic responses and adaptation for given long-term management strategies. This involves input from the social sciences (economics, social and political science, operational research) and from a wider range of natural sciences (ecology, genetics, oceanography) than traditionally is the case in fisheries management.

The committee has not explicitly considered the costs of implementing these recommendations, which may require either additional resources or a reprogramming of existing resources.

CENTRAL ISSUES

Adequacy of Scientific Information for Management

Improving Fish Stock Assessments (NRC, 1998) reviewed the state of existing knowledge about the stock assessment process. The report stressed that the feedback between stock assessment and fisheries management has to be improved to manage fisheries more effectively. The report also includes nine recommendations pertinent to stock assessment. The committee examined these recommendations in relation to the Northeast fishery stock assessments, and concluded that most of the earlier report's recommendations are already being addressed by NMFS in these fisheries. **The current stock assessment process, despite the need for improvements, appears to provide a valid scientific context for evaluating the status of fish populations and the effects of fishery management. Furthermore, the process is analogous to processes used in jurisdictions elsewhere in the world. Therefore, the Northeast stock assessment process is well within the standards of the stock assessments conducted elsewhere in the United States and by other nations.**

In all five stocks considered, fishing mortality was high, increasing and not sustainable, whereas spawning stock biomass was low and decreasing. The available data and the assessments using them show convincingly that the reviewed stocks have been subject to increased fishing mortality and decreasing spawning stock biomass through the 1980s and early 1990s. These conditions considerably increase the risk of major stock collapse. The increasing fishing mortality rates during the 1980s and 1990 indicate that management measures implemented during that period were ineffective in controlling fishing mortality. Therefore a different, more drastic approach was needed to decrease the probability of stock collapse.

Justification For Management Actions

Overall, the demonstrated tenuous status of the five stocks and the substantial uncertainty surrounding their ability to recover warranted strong management action. The committee finds no scientific basis to support assertions that the regulations imposed by Amendment 7 are too severe from a biological perspective. In fact, further management action may be necessary for the Gulf of Maine cod fishery. It should be added that the regulations in Amendments 5-7 might have been avoided if fishing mortality in the New England groundfisheries had been effectively controlled from the mid-1980s.

Conclusions

If stock sizes are at intermediate levels relative to historical values, an argument might be made simply to maximize the yield from recruitment obtained. However, current scientific wisdom indicates that a precautionary approach should be used for harvesting unless there is conclusive evidence that higher harvests will result in better ecosystem regulation. **When stocks are low, as in the Northeast, high fishing mortality markedly increases the risk of irrevocable stock damage.**

In two of the five cases in the Northeast (Georges Bank haddock and southern New England yellowtail flounder), the committee believes that the stocks already have collapsed, as indicated by low spawning levels combined with a period of little or no recruitment. Recruitment remains low and recent increases in biomass primarily result from higher survival of small year classes and growth of the current biomass because of the lower fishing mortality. For Georges Bank cod, yellowtail flounder, and Gulf of Maine cod, some of the committee's simulations of future stock size show that there would be a real danger of future stock collapse if strong regulations to reduce fishing mortality were not in place. For Gulf of Maine cod, the stock does not appear to have collapsed but there is danger it could under recent fishing mortality. Current regulations have not yet shown the ability to control fishing mortality for this stock. Additional management measures may be required.

Effects of Management Regulations

The assessments indicate that fishing mortality played a major role in reducing the abundance of groundfish in New England. Current stock assessments suggest that fishing mortality has been reduced for four of the five reviewed stocks and that these stocks appear to be increasing. The fifth stock, Gulf of Maine cod, has not experienced reduced fishing mortality, and it is not increasing. Biomass remains considerably smaller than observed in the past, and any relaxation of management measures may jeopardize sustained stock rebuilding.

Sufficiency of Survey and Landings Data

The scope and protocols of current data inputs (trawl survey and landing data) are sufficient for the stock assessments. Improvements in data collection systems are necessary, especially for catch and discard reporting, survey coverage, and collection of age information. Nevertheless, the assessment inputs are comparable to and, to some extent, better than those available for other stocks in the United States and elsewhere. The uncertainty of the stock assessments could be reduced further if the data inputs were improved.

Usefulness of Stock Assessments as Predictive Tools

There is no simple answer regarding the adequacy and reliability of these stock assessments as predictive tools. The current assessments evaluate how implementing different fishing mortalities (F), including no change in F, would affect the annual probabilities of reaching the rebuilding thresholds set by the New England Fishery Management Council. Future biomass trajectories can only be predicted in probabilistic terms, and there is great uncertainty about how long the stocks will take to rebuild to these thresholds.

While NMFS forecasts do incorporate substantial uncertainty, some of the main sources of uncertainty have been left out. Knowledge of the biology of the fish stocks is incomplete, and this contributes to uncertainty in stock projections. For example, it is unclear whether changes in recruitment of young fish are environmentally controlled or the result of low spawning stock sizes. **Different hypotheses about the relationship between stock abundance and subsequent recruitment are** consistent with the data at hand, and they tend to push the time of recovery further out into the future. Under some of these hypotheses, the probability of the stock not having recovered to the Council thresholds, say within 10 years, is thus substantially larger than suggested by NMFS assessments. To estimate such probabilities quantitatively is not easy. It is therefore difficult to base the management decisions directly on predictions from stock assessments, without embodying the assessment and its predictions in a management strategy with precautionary qualities.

Alternative Methods To Regulate Catch

Technical measures, such as mesh size increases, were used extensively in the Northeast groundfish fisheries during the 1980s and early 1990s, and they did not succeed in limiting fishing mortality. Mesh size restrictions often do not lead to conservation of fish stocks unless fishing mortality itself is somehow controlled, for example, through setting a Total Allowable Catch (TAC) for the species, establishing closed areas, or limiting fishing effort. Increases in the closed areas appear to have played an important role in reducing fishing mortality in 1995 and 1996. The New England Fishery Management Council may choose to adjust mesh size and closed areas to fine-tune the control of fishing mortality. One problem with so-called technical measures is that they make it harder to quantitatively determine the effects of management actions on fishing mortality and stock size. **A valuable role for future stock assessments will be to investigate whether and how the effects of these management actions on both fishing mortality and stock size can be measured.**

Factors Affecting Abundance of Stocks

Evaluating the relative contributions of different factors in driving changes in the abundance of fish stocks is difficult, but as mentioned above, fishing mortality played a major role in reducing the abundance of the five stocks in the 1980s and early 1990s, and still does for the Gulf of Maine cod stock. The most recent estimates of fish mortality are the most uncertain, and assessments in future years may come up with much different estimates for 1995 and 1996. It should be noted that hydroclimatic changes have occurred in U.S. waters of the northwestern Atlantic, but their magnitude is substantially less than off Newfoundland and Labrador, where they are believed to be one of the causative factors in stock collapses. **However, even if environmental changes were a factor affecting stock abundance, fishing mortalities were high enough to contribute to major stock collapses. To avoid stock collapse it is more important to stringently reduce fishing mortality during periods of adverse environmental conditions.**

Article 6.2 of the UN Agreement on Highly Migratory Fish Stocks and Straddling Fish Stocks (FAO, 1995) directs that: "States shall be more cautious when information is uncertain, unreliable or inadequate. The absence of adequate scientific information shall not be used as a reason for postponing or failing to take conservation and management measures." Further, Article 6.7 of this agreement specifies that: "If a natural phenomenon has a significant adverse impact on the status of [straddling fish stocks or highly migratory] fish stocks, States shall adopt conservation and management measures on an emergency basis to ensure that fishing activity does not exacerbate such adverse impact. States shall also adopt such measures on an emergency basis where fishing activity presents a serious threat to the sustainability of such stocks."

Other Causes of Stock Fluctuations

There is speculation among some interested parties that pollution and habitat destruction are reducing stock recruitment. If, in fact, these factors have affected recruitment (and there is

Conclusions

no consensus in the scientific community), then this would not alter the scientific recommendation to reduce harvest levels. The fishing mortality recommended for stock rebuilding would need to be adjusted downward for decreased recruitment. Furthermore, if these factors are likely to continue for a long time, then it may not be possible to rebuild stocks to levels which occurred in the past. The larger issue of establishing a causal mechanism between stock fluctuations and environmental and pollution factors is a difficult one because all factors are changing at the same time and so their effects are confounded. However, future directed research on this area, involving environmental and ecosystem studies and cross-population comparisons of key stock-recruitment parameters, should be very valuable for constructing likely scenarios for policy evaluation.

STOCK ASSESSMENT AND FISHERY MANAGEMENT

The committee concludes that stock assessment science is not the real source of contention in the management of New England groundfish fisheries. Comments at a public hearing held by the committee support this conclusion. Many speakers suggested that the social and economic concerns created by strong management measures and lack of participation in the management process were the more important concerns. Traditional fishery science has a major role to play in fisheries management, but sound stock assessment clearly is not the only consideration.

The New England Fishery Management Council will be facing critical decisions, depending on the recovery or nonrecovery of groundfish stocks. A long-term management strategy will be needed to decide the rate of rebuilding required to reach particular targets. Without sound stock assessment, targets and rebuilding rates cannot be set, nor can the effectiveness of the regulatory actions be measured. However, stock assessment in the narrow sense of estimating status and dynamics of fish populations is not sufficient for rational fisheries management.

What constitutes a good management approach will vary over time, location, and components of the fish stock. To obtain the information necessary to design effective institutional and regulatory frameworks, it is essential that management draws on stock assessment, oceanography, ecology, economics, social and political science and operational research. **Only when a more comprehensive approach is taken, with long-term management strategies based on data and insight from the various fields, properly accounting for the uncertainties surrounding data and theory, can fishery management provide for high continuing yield of food and health of stocks, while considering the needs of people dependent upon the fisheries.**

REFERENCES

Aguirre International. 1997. *An Appraisal of the Social and Cultural Aspects of the Multispecies Groundfishery in New England and the Mid-Atlantic Regions.* Report submitted to NOAA. Contract Number 50-DGNF-5-00008.

Ames, E. 1997. *Cod and Haddock Spawning Grounds in the Gulf of Maine.* Rockland, Maine: Island Institute.

Anonymous. 1992. *Techniques for Biological Assessment in Fisheries Management.* Report of the workshop Julich, July 17-24, 1991. Forschungszentrum Julich GmbH, Berichte aus der Okologishchen Forchung, Band 9.

Azarovitz, T.R. 1981. A brief historical review of the Woods Hole Laboratory trawl survey time series. Pp. 62-67 in W.G. Doubleday and D. Rivard (eds.), *Bottom Trawl Surveys.* Canadian Special Publication of Fisheries and Aquatic Sciences 58.

Brown, R. 1997. U.S.A. Assessment Appendices of the Georges Bank Haddock Stock, 1997. Working Paper, SAW-24. Northeast Fisheries Science Center Reference Document. unpublished manuscript.

Cadrin, S., W. Overholtz, J. Neilson, S. Gavaris, and S. Wigley. 1997. Stock Assessment of Georges Bank Yellowtail Flounder. Working Paper, SAW-24. Northeast Fisheries Science Center Reference Document. unpublished manuscript.

Chambers, R.C., and E.A. Trippel. 1997. *Early Life History and Recruitment in Fish Populations.* London: Chapman and Hall.

Clark, S.H., W.J. Overholtz, and R.C. Hennemuth. 1982. Review and assessment of the Georges Bank and Gulf of Maine haddock fishery. *J. Northwest Atl. Fish. Sci.* 3:1-27.

Clark, W.G. 1993. The effect of recruitment variability on the choice of a target level of spawning biomass per recruit. Pp. 233-246 in G. Kruse, D.M. Eggers, R. J. Marasco, C. Pautske, and T. Quinn II (eds.), *Proceedings of the International symposium on Management Strategies for Exploited Fish Populations*. Alaska Sea Grant College Program Report No. 93-02. University of Alaska, Fairbanks.

Clay, P., and E. Dolin. 1997. Building better social impact assessments. *Fisheries* 22:12-13.

Coe, R., and R.D. Stern. 1982. Fitting models to daily rainfall data. *J. App. Meteor.* 1024-1031.

Conser, R., and J.E. Powers. 1990. Extensions of the ADAPT VPA tuning method designed to facilitate assessment work on tuna and swordfish stocks. *ICCAT Coll. Vol. Sci. Pap.* 32:461-467.

Conser, R., R.D. Methot, and J.E. Powers. 1991. Integrative age/size-structured assessment methods: Stock synthesis, ADAPT and others. Working Paper, SAW-15. Northeast Fisheries Science Center Reference Document. unpublished manuscript.

DeLong, A., K. Sosebee, and S. Cadrin. 1997. Evaluation of vessel logbook data for discard and CPUE estimates. Working Paper, SAW-24, Northeast Fisheries Science Center Reference Document. unpublished manuscript.

Deriso, R.B. 1982. Relationship of fishing mortality to natural mortality and growth at the level of maximum sustainable yield. *Can. J. Fish Aquat. Sci.* 39:1054-1058.

Deriso, R.B., T.J. Quinn II, and P.R. Neal. 1985. Catch-age analysis with auxiliary information. *Can. J. Fish. Aquat. Sci.* 42:815-824.

Edwards, S.F., and S.A. Murawski. 1993. Potential economic benefits from efficient harvest of New England groundfish. *N. Am. J. Fish. Manage.* 13:437-449.

Food and Agriculture Organization (FAO). 1995. *Precautionary Approach to Fisheries*. FAO Fisheries Technical Report 350. Rome: United Nations.

Fordham, S. 1996. *New England Groundfish: From Glory to Grief. A Portrait of America's Most Devastated Fishery*. Washington, D.C.: Center for Marine Conservation.

Fournier, D.A., and C.P. Archibald. 1982. A general theory for analyzing catch at age data. *Can. J. Fish. Aquat. Sci.* 39:1195-1207.

Gavaris, S. 1991. Experience with the adaptive framework as a calibration tool for finfish stock assessment in CAFSAC. *ICES C.M.* 1991/D:19.

Gavaris, S. 1993. Analytical estimates of reliability for the projected yield from commercial fisheries. Pp. 185-191 in S.J. Smith, J.J. Hunt, and D. Rivard (eds.), *Risk Evaluation and Biological Reference Points for Fisheries Management*. Canadian Special Publication of Fisheries and Aquatics Sciences 120.

References

Gavaris, S., and L. VanEeckhaute. 1996. *Assessment of Haddock on Eastern Georges Bank*. Department of Fisheries and Oceans Atlantic Fisheries Research Document 96/21.

Gavaris, S., and L. Van Eeckhaute. 1997. *Assessment of Haddock on Eastern Georges Bank*. Department of Fisheries and Oceans Atlantic Fisheries Research Document 97/54.

Gavaris, S., J.J. Hunt, J.D. Neilson, and F. Page. 1996. *Assessment of Georges Bank Yellowtail Flounder*. Department of Fisheries and Oceans Atlantic Fisheries Research Document 96/22.

Grosslein, M.D. 1961. Haddock stock in the ICNAF Convention Area. *ICNAF Redbook 1962* Part III: 124-131.

Gunderson, D.R. 1993. *Surveys of Fisheries Resources*. New York: John Wiley and Sons.

Healey, M.C., and T. Hennessey. 1998a. The paradox of fairness: The impact of escalating complexity on fishery management. *Marine Policy* 22:109-118.

Healey, M.C., and T. Hennessey. 1998b. Ludwig's ratchet and the destruction of the New England groundfishery. *Marine Policy* (in press).

Hennemuth, R.C., and S. Rockwell. 1987. History of fisheries conservation and management. Pp. 430-436 in R.H. Baukus, (ed.), *Georges Bank*. Cambridge, Massachusetts: MIT Press.

Hilborn, R., and C.J. Walters. 1992. *Quantitative Fisheries Stock Assessment: Choice, Dynamics, and Uncertainty*. New York: Chapman and Hall.

Hunt, J.J., and M.I. Buzeta. 1996. *Biological Update of Georges Bank Cod in Unit Areas 5Zj,m for 1978-1995*. Department of Fisheries and Oceans Atlantic Fisheries Research Document 96/23.

Hunt, J.J., and M.I. Buzeta. 1997. *Population Status of Georges Bank Cod in Unit Areas 5Zj,m for 1978-1996*. Department of Fisheries and Oceans Atlantic Fisheries Research Document 97/33.

Hutchings, J., and R. Myers. 1994. What can be learned from the collapse of a renewable resource? Atlantic cod (*Gadus morhua*) off Newfoundland and Labrador. *Can. J. Fish. Aquat. Sci.* 51:2126-2146.

Hutchings, J., C. Walters, and R. Haedrich. 1997. Is scientific inquiry incompatible with government information control? *Can. J. Fish. Aquat. Sci.* 54:1198-1210.

Ianelli, J. 1997. An Alternative Assessment Analysis for Gulf of Maine Cod. Draft SARC Document A2. unpublished manuscript.

International Council for the Exploration of the Sea (ICES). 1993. *Report of the Working Group on Methods of Fish Stock Assessments*. ICES Cooperative Research Report 191.

Jensen, A. 1972. *The Cod*. New York: Thomas Y. Crowell Company.

Kurlansky, M. 1997. *Cod: A Biography of the Fish That Changed the World*. New York: Walker and

Company.

Lierman, M., and R. Hilborn. 1997. Depensation in fish stocks: a hierachic Bayesian meta-analysis. *Can. J. Fish. Aquat. Sci.* 54:1976-1983.

Mace, P.M., and M.P. Sissenwine. 1993. How much spawning biomass per recruit is enough? Pp. 101-118 in S.J. Smith, J.J. Hunt, and D. Rivard (eds.), *Risk Evaluation and Biological Reference Points for Fisheries Management.* Canadian Special Publication of Fisheries and Aquatic Sciences 120.

Marshall, C.T., and K.T. Frank. 1994. Do compensatory processes regulate haddock recruitment? *American Fisheries Society Annual Meeting Abstracts* 124:183-184.

Mayo, R.K. 1997. Assessment of the Gulf of Maine cod for 1997. Working Paper, SAW-24. Northeast Fisheries Science Center Reference Document. unpublished manuscript.

Mayo, R.K., M.J. Fogarty, and F.M. Serchuk. 1992. Aggregate fish biomass and yield on Georges Bank. *J. Northwest Atl. Fish. Sci.* 14:59-78.

McCallister, M.K., and J.N. Ianelli. 1997. Bayesian stock assessment using catch-age data and the sampling-importance resampling algorithm. *Can. J. Fish. Aquat. Sci.* 54:284-300.

McCracken, F.D. 1960. Studies of haddock in the Passamaquoddy Bay region. *J. Fish. Res. Bd. Canada* 17:175-180.

Methot, R.D. 1989. Synthetic estimates of historical abundance and mortality in northern anchovy. *Amer. Fish. Soc. Symp.* 6:66-82.

Multispecies Monitoring Committee (MMC). 1996. Report of the Multispecies Monitoring Committee to the New England Fishery Management Council. December.

Murawski, S.A., J.-J. Maguire, R.K. Mayo, and F.M. Serchuk. 1997. Groundfish stocks and the fishing industry. Pp. 25-68 in J.G. Boreman, B.S. Nakashima, and R.L. Kendall (eds.), *Northwest Atlantic Groundfish: Perspectives on a Fishery Collapse.* Bethesda, Maryland: American Fisheries Society.

Myers, R., and N. Cadigan. 1995. Was an increase in natural mortality responsible for the collapse of northern cod? *Can. J. Fish. Aquat. Sci.* 52:1274-1285.

Myers, R., J. Hutchings, and A. Rosenberg. 1995. Population dynamics of exploited fish stocks at low population levels. *Science* 269:1106-1108.

Myers, R., J. Hutchings, and N. Barrowman. 1996. Hypotheses for decline of cod in the North Atlantic. *Mar. Ecol. Prog. Ser.* 138:293-308.

Myers, R., J. Hutchings, and N. Barrowman. 1997. Why do fish stock collapse? The example of cod in Eastern Canada. *Ecol. Appl.* 7:91-106.

References

National Research Council (NRC). 1998. *Improving Fish Stock Assessments.* Washington, D.C.: National Academy Press.

New England Fishery Management Council (NEFMC). 1994. Amendment #5 to the Northeast Multispecies Fishery Management Plan.

New England Fishery Management Council (NEFMC). 1996. Amendment #7 to the Northeast Multispecies Fishery Management Plan.

Northeast Fisheries Science Center (NEFSC). 1994a. Report of the 18th Northeast Regional Stock Assessment Workshop (18th SAW). The Plenary. Northeast Fisheries Science Center Reference Document 94-23.

Northeast Fisheries Science Center (NEFSC). 1994b. Report of the 18th Northeast Regional Stock Assessment Workshop (18th SAW), December, 1994. Northeast Fisheries Science Center Reference Document 94-22.

Northeast Fisheries Science Center (NEFSC). 1994c. Report of the 17th Northeast Regional Stock Assessment Workshop (17th SAW), January, 1994. Northeast Fisheries Science Center Reference Document 94-06.

Northeast Fisheries Science Center (NEFSC). 1995. Report of the 19th Northeast Regional Stock Workshop (19th SAW). Northeast Fishery Science Center Reference Document 95-08.

Northeast Fisheries Science Center (NEFSC). 1997a. Report of the 24th Northeast Regional Stock Assessment Workshop (24th SAW): Stock Assessment Review Committee (SARC) Consensus Summary of Assessments.

Northeast Fisheries Science Center (NEFSC). 1997b. Report of the 24th Northeast Regional Stock Assessment Workshop (24th SAW): Stock Assessment Review Committee (SARC) Advisory Report on Stock Status.

Neilson, J., S. Gavaris, and J. Hunt. 1997. *Assessment of Georges Bank (5Zjmnh) Yellowtail Flounder* (Limanda ferruginea). Department of Fisheries and Oceans Atlantic Fisheries Research Document 97/55. unpublished manuscript.

O'Brien, L. 1997. Assessment of the Georges Bank Haddock Stock. Working Paper, SAW-24. Northeast Fisheries Science Center Reference Document. unpublished manuscript.

O'Brien, L., and R. Brown. 1995. Assessment of the Georges Bank haddock stock for 1994. Northeast Fisheries Science Center Reference Document 95-13.

Organization for Economic Co-operation and Development (OECD). 1997. *Towards Sustainable Fisheries: Economic Aspects of the Management of Living Marine Resources.* Paris: OECD Press.

Overholtz, W., S. Cadrin, and S. Wigley. 1997. Assessment of the Southern New England Yellowtail Flounder Stock for 1997. Working Paper SAW-24. Northeast Fisheries Science Center Reference

Document. unpublished manuscript.

Overholtz, W.S., S.F. Edwards, and J.K.T. Brodziak. 1993. *Strategies for rebuilding and harvesting New England Groundfish Resources.* Pp. 507-527 in G. Kruse, D. Eggers, R. Marasco, C. Pautzke, and T. Quinn II (eds.), Proceedings of the International Symposium on Management Strategies for Exploited Fish Populations. Alaska Sea Grant College programs Report. No. 93-02. Fairbanks: University of Alaska.

Pálsson, Ó.K., E. Jónsson, S.A. Schopka, G. Stefánsson, and B. Steinarsson. 1989. Icelandic groundfish survey data used to improve precision in stock assessments. *J. Northwest Atl. Fish. Sci.* 9:53-72.

Patterson, K.R., and G.P. Kirkwood. 1993. Comparative performance of ADAPT and Laurec-Shepherd methods for estimating fish population parameters and in stock management. *ICES J. Mar. Sci.* 52:183-196.

Royce, W.F., R.J. Buller, and E.D. Premetz. 1959. Decline of the yellowtail flounder (*Limanda ferruginea*) off New England. *Fish. Bull. (U.S.)* 59:169-267.

Serchuk, F.M., and S.E. Wigley. 1992. Assessment and management of the Georges Bank cod fishery: An historical review and evaluation. *J. Northwest Atl. Fish. Sci.* 13:25-52.

Serchuk, F.M., M.D. Grosslein, R.G. Lough, D.G. Mountain, and L. O'Brien. 1994. Fishery and environmental factors affecting trends and fluctuations in the Georges Bank and Gulf of Maine Atlantic cod stocks: An overview. *ICES Mar. Sci. Symp.* 198:77-109.

Thompson, S.K., and G.A.F. Seber. 1996. *Adaptive Sampling.* New York: Wiley.

APPENDIXES

Appendix A

Mandate from Magnuson-Stevens Fishery Conservation and Management Act

Authorizing Legislation:

Magnuson-Stevens Fishery Conservation and Management Act (16 U.S.C. 1801 et seq.):

SEC. 210. REVIEW OF NORTHEAST FISHERY STOCK ASSESSMENTS.

The National Academy of Sciences, in consultation with regionally recognized fishery experts, shall conduct a peer review of Canadian and U.S. stock assessments, information collection methodologies, biological assumptions and projections, and other relevant scientific information used as the basis for conservation and management in the Northeast multispecies fishery. The National Academy of Sciences shall submit the results of such review to the Congress and the Secretary of Commerce no later than March 1, 1997.*

*Due to the timing of the 1997 stock assessment process (Figure 1.5), and delays in funding, it was agreed that the deadline for the completion of the report would be extended.

Appendix B

Committee Biographies

Terrance J. Quinn II (*chairman*) earned his Ph.D. in biomathematics from the University of Washington in 1980. Dr. Quinn has been an associate professor at the University of Alaska since 1985. He is a member of the Scientific and Statistical Committee of the North Pacific Fisheries Management Council and has recently served as consultant to the National Marine Fisheries Service (NMFS) and the Makah Indian Tribe in Neah Bay, Washington. Dr. Quinn also served on the Ocean Studies Board (OSB) Committee on Fisheries and Committee to Review Atlantic Bluefin Tuna; he co-chaired the Committee on Fish Stock Assessments and was appointed to the OSB in 1995. His research interests are in the areas of fish population dynamics and management and applied statistics and biometrics.

Wyatt Anderson earned his Ph.D. from Rockefeller University in 1967. He has been a Distinguished Professor of Genetics at the University of Georgia since 1988. Dr. Anderson is a member of the National Academy of Sciences. His research interests include molecular evolution, genetics of natural and experimental populations, theoretical population genetics, scientific literacy, and scientific education.

Wayne M. Getz earned his Ph.D. in applied mathematics from the University of the Witwatersrand, Johannesburg, South Africa in 1976. He currently is professor of environmental sciences and biomathematician in the Agricultural Experiment Station at the University of California, Berkeley, where he joined the faculty in 1979. Dr. Getz's research interests include population modeling, resource management, and self-organizing systems.

Ray Hilborn earned his Ph.D. in zoology from the University of British Columbia in 1974. Dr. Hilborn is professor at the School of Fisheries of the University of Washington. His main areas of research are resource management, population dynamics, systems analysis, and fisheries.

Cynthia Jones earned her Ph.D. in oceanography from the University of Rhode Island in 1984. She is an associate professor in the Department of Biology at Old Dominion University. Dr. Jones' research focuses on fisheries and population ecology.

Jean-Jacques Maguire earned his M.Sc. in biology from Laval University in 1984. He worked for the Canadian Department of Fish and Oceans from 1977 to 1996. Mr. Maguire is now a private consultant in fisheries science and management and is the chairman of the Advisory Committee on Fisheries Management for the International Council on the Exploration of the Sea. His expertise is in the area of groundfish and large pelagic fish stock management.

Appendix B

Ana Parma earned her Ph.D. in fisheries from the University of Washington in 1988. She is currently a population dynamicist for the International Pacific Halibut Commission. Her research focuses on population dynamics and adaptive fisheries management.

Tore Schweder earned his Ph.D. in statistics from the University of California, Berkeley in 1974. He has been a professor of statistics at the University of Oslo since 1984 and is currently on sabbatical leave at Stanford University. Dr. Schweder's areas of research include statistical methodology, demography, population biology, line transect surveys, and medicine. Dr. Schweder has developed a multispecies fisheries model including both biological and economic factors.

Gunner Stefansson earned his Ph.D. in statistics from Ohio State University in 1983. He has been a manager in the Modeling Division of the Marine Research Institute, Iceland, since 1983. Dr. Stefansson's research interests include multiple comparisons in statistics and survey data used to improve precision in stock assessments.

# Appendix C

Materials Received from National Marine Fisheries Service (NMFS)

Data and Background Information on Previous Stock Assessments
Received by NRC: April 3, 1997.

I. DATA SOURCES AND ANALYSIS METHODS: ASSESSMENT DATA SOURCES AND PROTOCOLS

(1) Research Vessel Surveys

a) NMFS Offshore Surveys:

I-1   Arazovitz, T.R. 1989. Northeast Fisheries Center trawl surveys. Pp. 19-22 in T.R. Arazovitz and J. McGurrin (eds.). Proceedings of a Workshop on Bottom Trawl Surveys. Special Report 17. Atlantic States Marine Fisheries Commission.

I-2   Byrne, C.J., T.R. Arazovitz, and M.P. Sissenwine. 1981. Factors affecting variability of research vessel trawl surveys. Pp. 258-273 in W.G. Doubleday and D. Rivard (eds.), *Bottom Trawl Surveys*. Canadian Special Publication of Fisheries and Aquatic Sciences 58.

I-3   Arazovitz, T.R. 1981. A brief historical review of the Woods Hole Laboratory trawl survey time series. Pp. 62-67 in W.G. Doubleday and D. Rivard (eds.), *Bottom Trawl Surveys*. Canadian Special Publication of Fisheries and Aquatic Sciences 58.

I-4   Grosslein, M.D. 1969. Groundfish survey program at BCF Woods Hole. *Commercial Fish. Rev.* 3198-3199:22-30.

I-5   Survey data:
       (1) Survey Strata Maps and Areas of Strata
       (2) Spring Survey History
       (3) Autumn Survey History

b) State of Massachusetts Inshore Surveys:

I-6   Howe, A.B. 1989. State of Massachusetts inshore bottom trawl survey. Pp. 33-38 in T.R. Arazovitz and J. McGurrin (eds.), *Proceedings of a Workshop on Bottom Trawl Surveys.* Special Report 17. Atlantic States Marine Fisheries Commission.

c) Canadian Surveys:

I-7   Description of time series and some summary data from the 1996 winter survey

(2)   Commercial Landings Data

a) Pre-1994 U.S. Commercial Data (landings and biological sampling):

I-8   Burns, T.S., R. Schultz, and B.E. Brown. 1983. The commercial catch sampling program in the northeastern United States. Pp. 82-95 in W.G. Doubleday and D. Rivard (eds.), *Bottom Trawl Surveys.* Canadian Special Publication of Fisheries and Aquatic Sciences 58.

b) 1994-Present Commercial Data:

I-9   Descriptions of the current mandatory reporting program, as well as data fields and handling procedures.
       (1) Dealer Reports
       (2) Vessel Reports

An evaluation of vessel data collection and handling procedures under the new mandatory program was undertaken at SARC 22. That evaluation is contained in the following:

I-10   Analysis of 1994 fishing vessel logbook data. Pp. 8-46 in Report of the 22nd Northeast Regional Stock Assessment Workshop (22nd SAW). Consensus summary of assessment. Northeast Fisheries Science Center Reference Document 96-13.

c) Biological Sampling of Landings:

I-11   Tables of length and age sampling requirements currently used by port samplers to obtain biological information for cod, haddock, and yellowtail flounder.

(3)   Recreational Fisheries Data

I-12   Van Voorhees, D.A., J.F. Witzig, M.F. Osborn, M.C. Holliday, and R.J. Essig. 1992. *Marine Recreational Fishery Statistics Survey, Atlantic and Gulf Coasts*, 1990-1991. National Marine Fisheries Service. Current Fisheries Statistics Number 9204. Silver Spring, MD.

(4)   Sea Sampling (Observer) Data

I-13   *Northeast Fisheries Science Center Observer Manual*

I-14   Murawski, S.A. 1996. Factors influencing by-catch and discard rates: Analyses from multispecies/multifishery sea sampling. *J. Northwest Atl. Fish. Sci.* 19:31-39.

I-15 Murawski, S.A., K. Mays, and D. Christensen. 1994. Fishery observer program. In *Status of the Fishery Resources off the Northeastern United States for 1994.* NOAA Technical Memorandum NMFS-NE-108.

I-16 Hayes, D. 1991. *Exploratory Analysis of Four Methods for Estimating Discards from Sea Sampling Data.* Northeast Fisheries Science Center CRD-91-03.

Stock Assessment Methods and Procedures

(5) Research Vessel Abundance Indices

I-17 Northeast Demersal Complex. Pp. 134-196 in Report of the 21st SAW. Consensus summary of assessments. June 1996. Northeast Fisheries Science Center Reference Document 96-05d.

I-18 Ameida, F.P., M.J. Fogarty, S.H. Clark, and J.S. Idoine. 1986. An evaluation of precision of abundance estimators derived from bottom trawl surveys off the northeastern United States. ICES C.M. 1986/G:91.

I-19 Pennington, M., and B.E. Brown. 1981. Abundance estimators based on stratified random trawl surveys. Pp. 149-153 in W.G. Doubleday and D. Rivard (eds.), *Bottom Trawl Surveys.* Canadian Special Publication of Fisheries and Aquatic Sciences 58.

(6) CPUE Measures

I-20 Mayo, R.K., T.E. Helser, L. O'Brien, K.A, Sosbee, B.F. Figurido, and D. Hayes. 1994. Estimation of standardized otter trawl effort, landings per unit effort, and landings at age for Gulf of Maine and Georges Bank cod. Northeast Fisheries Science Center Reference Document 94-12. 17 pp.

I-21 Mayo, R.K., M.J. Fogarty, and F.M. Serchuk. 1992. Aggregate fish biomass and yield on Georges Bank. *J. of Northwest Atl. Fish. Sci.* 14:59-78.

I-22 Gavaris, S. 1980. Use of a multiplicative model to estimate catch rate and effort from commercial data. *Canadian Journal of Fish. Aquat. Sci.* 37:2272-2275.

(7) Catch-at-Age Analysis

a) ADAPT Tuning:

I-23 Conser, R. 1993. A brief history of ADAPT. Working Document for SAW-15. Northeast Fisheries Science Center. unpublished manuscript.

I-24 Conser, R., R.D. Methot, and J.E. Powers. 1991. Integrative age/size-structured assessment methods: Stock synthesis, ADAPT and others. Working paper prepared for NMFS Workshop

Appendix C

on Stock Assessment Methods, March.

I-25 Gavaris, S. 1991. Experience with the adaptive framework as a calibration tool for finfish stock assessment in CAFSAC. ICES C.M. 1991/D:19.

I-26 Conser, R., and J.E. Powers. 1989. Extensions of the ADAPT VPA tuning method designed to facilitate assessment work on tuna and swordfish stocks. ICCAT Working Document SCRS/89/43.

I-27 Gavaris, S. 1988. An adaptive framework for the estimation of population size. CAFSAC Research Document 88/29.

b) Retrospective Analysis of Assessment Performance:

I-28 Bias in SARC assessment results. Pp. 178-188 in Report of the 18th SAW, December, 1994. Northeast Fisheries Science Center Reference Document 94-22.

I-29 Report of the Assessment Methods Subcommittee, May 1994. Northeast Fisheries Science Center. Draft report.

(8) Projections

a) Short Term (1-3 Years Ahead):

See I-29 above.

b) Medium and Long Term (used for Amendments 5 and 7 of the Northeast Multispecies FMP):

I-30 Brodziak, J.K.T., and P.J. Rago. 1996. Agrepo Users Guide Version 1.1. Northwest Fisheries Science Center, Newport, Ore. unpublished manuscript.

I-31 Amendment 7 to the Northeast Multispecies FMP. Appendix VII. Projections of landings, spawning stock biomass and recruitment under the status quo and Amendment 7 for Georges Bank and Gulf of Maine cod, Georges Bank haddock, and Georges Bank and southern New England yellowtail flounder.

II. MOST RECENT STOCK ASSESSMENTS

(1) U.S. Assessments (1994-1995)

a) Georges Bank Cod (SARC 18):

II-1 Georges Bank Cod. Pp. 155-177 in Report of the 18th Northeast Regional Stock Assessment Workshop (18th SAW), December 1994. Northeast Fisheries Science Center Reference Document 94-22.

II-2 Serchuk, F.R. Mayo, and L. O'Brien. 1994. Assessment of Georges Bank cod stock for 1994. Northeast Fisheries Science Center Reference Document 94-25 (includes expanded information and diagnostics of ADAPT tuning fit).

b) Gulf of Maine Cod (SARC 19):

II-3 Gulf of Maine Cod. Pp. 13-55 in Report of the 19th Northeast Regional Stock Assessment Workshop (19th SAW), 1995. Northeast Fisheries Science Center Reference Document 95-08.

II-4 Mayo, R. K. 1995. Assessment of the Gulf of Maine cod stock for 1994. Northeast Fisheries Science Center Reference Document 95-02 (includes expanded information and diagnostics of ADAPT tuning fit).

c) Georges Bank Haddock (SARC 20):

II-5 Georges Bank Haddock. Pp. 8-29 in Report of the 20th Northeast Regional Stock Assessment Workshop (20th SAW), February 1996. Northeast Fisheries Science Center Reference Document 95-18.

II-6 O'Brien, L., and R. Brown. 1995. Assessment of the Georges Bank haddock stock for 1994. Northeast Fisheries Science Center Reference Document 95-13 (includes expanded information and diagnostics of ADAPT tuning fit).

d) Georges Bank Yellowtail Flounder (SARC 18):

II-7 Georges Bank Yellowtail Flounder. Pp. 135-154 in Report of the 18th Northeast Regional Stock Assessment Work shop (18th SAW), December 1994. Northeast Fisheries Science Center Reference Document 94-22.

e) Southern New England Yellowtail Flounder (SARC 17):

II-8 Southern New England yellowtail Flounder. Pp. 36-54 in Report of the 17th Northeast Regional Stock Assessment Workshop (17th SAW), January 1994. Northeast Fisheries Science Center Reference Document 94-06.

(2) Canadian Assessment (1996)

a) Georges Bank Cod:

II-9 Hunt, J.J. and M.I. Buzeta. 1996. Biological update of Georges Bank cod in unit areas 5Z j,m for 1978-1995. Department of Fisheries and Oceans Atlantic Fisheries Research Document 96/23.

b) Georges Bank Haddock:

II-10 Gavaris, S., and L. Van Eeckhaute. 1996. Assessment of haddock on eastern Georges Bank. Department of Fisheries and Oceans Atlantic Fisheries Research Document 96/21.

Appendix C

c) Georges Bank Yellowtail Flounder:

II-11 Gavaris, S., J. J. Hunt, J. D. Neilson, and F. Page. 1996. Assessment of Georges Bank yellowtail flounder. Department of Fisheries and Oceans Atlantic Fisheries Research Document 96/22.

III. MANAGEMENT ADVICE

(1) Stock Assessment Workshop Management Advice for New England Groundfish (1993-1995)

a) Georges Bank Cod:

III-1 Georges Bank Cod Advisory Report. Pp. 47-50 in Report of the 18th Northeast Regional Stock Assessment Workshop (18th SAW). The Plenary. Northeast Fisheries Science Center Reference Document 94-23.

b) Gulf of Maine Cod:

III-2 Gulf of Maine Cod (Division 5Y) Advisory Report. Pp. 21-28 in Report of the 19th Northeast Regional Stock Assessment Workshop (19th SAW), March 1995. The Plenary. Northeast Fisheries Science Center Reference Document 95-09.

c) Georges Bank Haddock:

III-3 Georges Bank Haddock (Division 5Z and South) Advisory Report. Pp. 9-14 in Report of the 20th Northeast Regional Stock Assessment Workshop (20th SAW), May 1996. SAW Public Review Workshop. Northeast Fisheries Science Center Reference Document 95-19.

d) Georges Bank Yellowtail Flounder

III-4 Georges Bank Yellowtail Flounder Advisory Report. Pp. 42-46 in Report of the 18th Northeast Regional Stock Assessment Workshop (18th SAW). The Plenary. Northeast Fisheries Science Center Reference Document 94-23.

e) Southern New England Yellowtail Flounder:

III-5 Southern New England Yellowtail Flounder. Pp. 27-31 in Report of the 17th Northeast Regional Stock Assessment Workshop (17th SAW). The Plenary, March 1994. Northeast Fisheries Science Center Reference Document 94-07.

f) Groundfish Status on Georges Bank:

III-6 Special Advisory: Groundfish Status on Georges Bank. Pp. 53-57 in Report of the 18th Northeast Regional Stock Assessment Workshop (18th SAW). The Plenary. Northeast Fisheries Science Center Reference Document 94-23.

(2) Multispecies Monitoring Committee (New England Fishery Management Council)

- Evaluation and Advice (1996):

III-7 Report of the Multispecies Monitoring Committee to the New England Fishery Management Council. December 1996.

(3) Department of Fisheries and Oceans (DFO) Stock Status Summaries

- 1994:

III-8 Report on the Status of Groundfish Stocks in the Canadian Northwest Atlantic. Department of Fisheries and Oceans Stock Status Report 94/4 (portions relevant to Georges Bank stocks).

- 1995:

III-9 Scotia-Fundy Spring 1995 Groundfish Status Report. DFO Atlantic Fisheries Stock Status Report 95/6. Dartmouth, N.S., Canada (Overview sections of the report plus relevant sections on Georges Bank cod, haddock, and yellowtail flounder).

- 1996:

III-10 Overview of the status of Canadian managed groundfish stocks in the Gulf of St. Lawrence and in the Canadian Atlantic. Atlantic Stock Assessment Secretariat, DFO, Ottawa. Stock Status Report 96/40E.

III-11 Georges Bank Cod. DFO Atlantic Fisheries Stock Status Report 96/21/E.

III-12 Eastern Georges Bank Haddock. DFO Atlantic Fisheries Stock Status Report 96/19/E.

III-13 Yellowtail Flounder on Georges Bank. DFO Atlantic Fisheries Stock Status Report 96/20/E.

(4) Fisheries Resource Conservation Council (Canada) Management Advice (1994-1996)

III-14 Fisheries Resource Conservation Council. 1994. Conservation Stay the Course. 1995 Conservation requirements for Atlantic groundfish. Report to the Minister of Fisheries and Oceans. FRCC.94.R.4E. (Portions relevant to Georges Bank stocks are reproduced).

III-15 Fisheries Resource Conservation Council. 1995. Conservation Come Aboard. 1996 Conservation requirements for Atlantic groundfish. Report to the Minister of Fisheries and Oceans. FRCC.94.R.2. (Portions relevant to Georges Bank stocks are reproduced).

III-16 Fisheries Resource Conservation Council. 1996. Building the Bridge. 1997 Conservation requirements for Atlantic groundfish. Report to the Minister of Fisheries and Oceans. FRCC.96.R.2. (Portions relevant to Georges Bank stocks are reproduced).

Appendix C

IV. BACKGROUND PAPERS

(1) Atlantic Cod

IV-1 Serchuk, F.M., and S.E. Wigley. 1992 Assessment and management of the Georges Bank cod fishery: An historical review and evaluation. *J. of Northwest Atl. Fish. Sci.* 13:25-52.

IV-2 Serchuk, F.M., M.D. Grosslein, R.G. Lough, D.G. Mountain, and L. O'Brien. 1994. Fishery and environmental factors affecting trends and fluctuations in the Georges Bank and Gulf of Maine Atlantic cod stocks: An overview. *ICES Mar. Sci. Symp.* 198:77-109.

IV-3 Penttila, J.A., and V.M. Gifford. 1976. Growth and mortality rates of cod from the Georges Bank and Gulf of Maine areas. *ICNAF Res. Bull.* 12:29-36.

(2) Haddock

IV-4 Clark, S.H., W.J. Overholtz, and R.C. Hennemuth. 1982. Review and assessment of the Georges Bank and Gulf of Maine haddock fishery. *J. of Northwest Atl. Fish. Sci.* 3:1-27.

IV-5 Overholtz, W.J., M.P. Sissenwine, and S.H. Clark. 1986. Recruitment variability and its implication for managing and rebuilding the Georges Bank haddock (*Melanogrammus aeglefinus*) stock. *Can. J. of Fish. and Aquat. Sci.* 43: 748-753.

IV-6 Halliday, R.G. 1988. Use of seasonal spawning area closures in the management of haddock fisheries in the northwest Atlantic. *NAFO Sci. Council Stud.* 12:27-36.

(3) Yellowtail Flounder

IV-7 Royce, W.F., R.J. Buller, and E.D. Premetz. 1959. Decline of the yellowtail flounder (*Limanda ferruginea*) off New England. *Fish. Bull. (U.S.)* 59:169-267.

IV-8 Lux, F.E. 1963. Identification of New England yellowtail flounder groups. *Fish. Bull. (U.S.)* 63(1):1-10.

IV-9 Lux, F.E. 1969. Landings per unit of effort, age composition, and total mortality of yellowtail flounder, *Limanda ferruginea* (Storer), off New England. *ICNAF Res. Bull.* 6:47-52.

IV-10 Lux, F.E., and F.E. Nichy. 1969. Growth of yellowtail flounder, *Limanda ferruginea* (Storer), on three New England fishing grounds. *ICNAF Res. Bull.* 6:5-25.

IV-11 Sissenwine, M.P. 1974. Variability in recruitment and equilibrium catch of the Southern New England yellowtail flounder fishery. *J. Cons. Int. Explor. Mer* 36(1):15-26.

(4) Generic Issues

- Maturity:

IV-12 O'Brien, L., J. Burnett, and R.K. Mayo. 1993. Maturation of nineteen species of finfish off the Northeast coast of the United States, 1985-1990. NOAA Technical report NMFS 113,

IV-13 Hunt, J.J. 1996. Rates of sexual maturation of Atlantic cod in NAFO Division 5Ze and commercial fishery implications. *Journal of Northwest Atlantic Fishery Science* 18:61-75.

- Age and Growth:

IV-14 Penttila, J.A., G.A. Nelson, and J.M. Burnett III. 1989. Guidelines for estimating lengths at age for 18 Northwest Atlantic finfish and shellfish species. NOAA Technical Memorandum NMFS-F/NEC-66.

[See also detailed aging procedures for each species contained on the Fishery Biology Investigation homepage at: http://www.wh.whoi.edu/fbi/age-man.html]

- Vessel, Door, and Net Calibration Coefficients:

IV-15 Byrne, C.J., and J.R.S. Forrester. 1991. Relative fishing power of NOAA R/V's Albatross IV and Delaware II. Working Paper, SARC 12. Northeast Fisheries Science Center. unpublished manuscript.

IV-16 Byrne, C.J., and J.R.S. Forrester. 1991. Relative fishing power two types of trawl doors. Working Paper, SARC 12. Northeast Fisheries Science Center. unpublished manuscript.

IV-17 Forrester, J.R.S. A trawl survey conversion coefficient suitable for lognormal data. *Biometrics* (in press). unpublished manuscript.

IV-18 Sissenwine, M.P. and E.W. Bowman. 1978. An analysis of some factors affecting the catchability of fish by bottom trawls. *ICNAF Res. Bull.* 13:81-87.

- Evaluation of NMFS Offshore Survey Program:

IV-19 Survey Working Group, Northeast Fisheries Center. 1988. An evaluation of the bottom trawl survey program of the Northeast Fisheries Center. NOAA Technical Memorandum NMFS-F/NEC-52.

Draft SARC Documents and Data Collection Documents Received by NRC: May 12, 1997

- Gulf of Maine Cod:

A1 Mayo, R.K. 1997. Assessment of the Gulf of Maine Cod for 1997.

A2   Ianelli, J. 1997. An Alternative Stock Assessment Analysis for Gulf of Maine Cod.

- Georges Bank Cod:

B1   O'Brien, L. 1997. Assessment of the Georges Bank Cod Stock, 1997.
Georges Bank Haddock:

C1   Brown, R. U.S.A. 1997. Assessment Appendices of the Georges Bank Haddock Stock, 1997.

- Georges Bank Yellowtail Flounder:

D1   S. Cadrin, W. Overholtz, and S. Wigley. 1997. Stock Assessment of Georges Bank Yellowtail Flounder.

D2   Almeida, F., and J. Burnett. 1997. Changes in Growth and Maturation of Yellowtail Flounder, *Pleuronectes ferrungineus*, from the Southern New England and Georges Bank Stocks During Periods of High and Low Abundance.

- Southern New England Yellowtail Flounder:

E1   Overholtz, W., Cadrin, S., and S. Wigley. 1997. Assessment of the Southern New England Yellowtail Flounder Stock for 1997.

All Stocks

Gen 1 Report of the Northern Demersal and Southern Demersal Working Groups. 1997.

Gen 2 Overholtz, S. 1997. Ten-Year Projections of Landings, Spawning Stock Biomass, and Recruitment for the Five Groundfish Stocks Considered at SAW-24.

Gen 3 Power, G., Wilhelm, K., McGrath, K., and T. Theriault. 1997. Commercial Fisheries Dependent Data Collection in the Northeastern United States.

Gen 4 Wigley, S., Terceiro, M., DeLong, A., and K. Sosebee. 1997. Proration of 1994-1996 Commercial Landings of Cod, Haddock, and Yellowtail Flounder.

Gen 5 DeLong, A., Sosebee, K., and S. Cadrin. 1997. Evaluation of Vessel Logbook Data for Discard and CPUE Estimates.

Gen 6 Forrester, J. et al. Background Papers on U.S.A. Vessel, Trawl, and Door Conversion Studies.

**Final SARC Documents
Received by NRC: June 10, 1997**

1   Report of the 24th Northeast Regional Stock Assessment Workshop (24th SAW). Stock Assessment Review Committee (SARC) Consensus Summary of Assessments.

2   Report of the 24th Northeast Regional Stock Assessment Workshop (24th SAW). Advisory Report on Stock Status.

Background ADAPT Documents and APL Workspace Diskettes
Received by NRC: June 20, 1997

1   Brodziak, J.K.T., and P.J. Rago. 1997. Agrepo's User's Guide. NWFSC/NEFSC.

2   Georges Bank Haddock Assessment SAW/SARC 24 Data.

3   Southern New England Yellowtail Flounder Assessment SAW/SARC 24 Data.

4   Georges Bank Yellowtail Flounder Assessment SAW/SARC 24 Data

5   Gulf of Maine Cod Assessment SAW/SARC 24 Data.

6   Georges Bank Cod Assessment SAW/SARC 24 Data.

7   Projection Source Code.

9   Diskettes of ADAPT APL workspaces as used for Assessments 2-7 above.

10   Diskettes of Excel spreadsheets for the time survey data for stocks 2-7 above.

11   Diskettes for source code and example output for prediction programs used for the stocks 2-7 above.

Background Commercial Fishery Data Documents
Received by NRC: June 24, 1997

1   Buzeta, M.I., J. J. Hunt, L. VanEeckhaute, and N. Munroe. 1991. Georges Bank Cod and Haddock Aging Workshop: September 10-13 St. Andrews, N.B. CAFSAC 92/119.

2   Fishery Biology Investigation: Program Review. 1989. Population Biology Branch, Conservation and Utilization Division. NEFSC.

3   Report of the Fishery Biology Investigation: Program Briefing. 1989. Population Biology Branch, Conservation and Utilization Division. NEFSC.

4   VanEeckhaute, L., M.I. Buzeta, N. Munroe, and V. Silva. 1994. Georges Bank Cod and Haddock Aging Exchange and Workshop. CAFSAC 94/84.

5   Northeast Regional Stock Assessment Workshops. 1994. NEFSC.

Appendix C

6   Northeast Marine Fisheries Information System (NEMFIS). 1994. Commercial Data Entry Subsystem (CODES). Operation Manual, Version 5.2. NEFSC/DMS.

7   Sea Sampling Data Entry System (SEASAMP). 1997. Operations Manual. NEFSC/DMS.

8   SVDBS Auditing Manual. Version 1.01. Prepared by Paul Kostovick.

**Background Recreational Fishery Data Documents
Received by NRC: June 27, 1997**

1   National Marine Recreational Fishery Statistics Survey. NOAA/NMFS. Solicitation Number 52-DGNF-5-00079.

2   National Marine Recreational Fishery Statistics Survey. NOAA/NMFS. Appendix A: Marine Recreational Fishery Statistics Survey Procedural Manual. Solicitation Number 52-DGNF-5-00079.

Appendix D

Presentation to the Committee by NMFS Scientists
(9 July, 1997, Bedford, Massachusetts)

Agenda

Time	Issue	Presenter
10:00-10:10	Overview, Introductions	M. Sissenwine
10:10-10:20	Stock Assessment Workshop Process, Terms of Reference, Interaction with Canadian Process	E. Anderson
10:20-10:45	Generic Issues: Databases, Survey Calibration, Assessment, and Prediction Methods	S. Murawski
	Assessments Inputs, Assumptions, Results, and Research Recommendations	S. Murawski
10:45-11:00	Georges Bank Haddock	R. Mayo
11:00-11:15	Georges Bank Cod	R. Mayo
11:15-11:30	Gulf of Maine Cod	R. Mayo
11:30-11:45	Southern New England Yellowtail Flounder	W. Overholtz
11:45-12:00	Georges Bank Yellowtail Flounder	W. Overholtz
12:00-12:45	*Lunch*	
12:45-1:15	Management Advice, Context with Previous Assessments, A#7/A#5, Canadian Management	M. Sissenwine
1:15-1:45	Questions and Discussion	

Information Presented

NORTHEAST REGIONAL STOCK ASSESSMENT WORKSHOP (SAW PROCESS)

- Steering Committee
- Working Groups
- Stock Assessment Review Committee (SARC)
- Public Review Workshop

SAW-24 AGENDA

- **Stocks**
  - Gulf of Maine Cod
  - Georges Bank Cod
  - Georges Bank Haddock
  - Georges Bank Yellowtail Flounder
  - Southern New England Yellowtail Flounder
- **Terms of Reference**
  - Assess the stock status through 1996 and characterize the variability of estimates of stock abundance and fishing mortality rates
  - Provide projected estimates of catch for 1997-1998 and SSB for 1998-1999 at various level of F, including all relevant biological reference points
  - Advise on the assessment and management implications of incorporating recreational catch and commercial discard data in the assessment

Canadian Data and Assessments

[Timeline diagram showing Canadian and United States Data and Assessments activities from Feb 1 through Jun 1, including:]

Canadian items (above timeline):
- '96 Commercial Ages week of 2/10
- '96 Landings Complete 2/17
- '96 Catch-at-Age Early March
- '97 Survey without ages 3/14
- '97 Survey with ages 3/21
- Prelim. ADAPT Runs week of 3/25
- Participate in ND Subcommittee
- RAP Moncton, N.B. Canada 4/21-25
- Internal DFO Review
- Submit Assessments to FRCC 5/7
- RAP = Regional Assessment Process

United States items (below timeline):
- '96 Survey Ages 2/5
- '96 Comm Ages 2/5
- '96 Fall Survey 2/15
- '96 Dealer 2/15
- Logbook 2/28
- '96 Sea Sampling
- '97 Spring Survey 2/24-3/7
- '96 Final Rec Data 3/17
- '96 Catch-at-Age Est. 3/21
- Prelim. ADAPT Runs 3/28
- ND Subcommittee 4/1-4/11 Woods Hole
- Participate in RAP and Conduct 10-Yr Projections by 4/30
- Documents Final by 5/9
- Assessments Distributed to SARC 5/9-10
- Data and Methods Description to NRC 5/9
- SARC Peer Review 5/19-23
- Final Assessments Distributed to NRC 6/7
- SAW Public Review Meeting 7/10
- ND = Northern Demersal Sub-Committee

United States Data and Assessments

Generic Assessment Issues I

- **Assessment Databases**
  - Changes in Landings Data Collection System
  - Proration of Landings to Stock Area
  - Effort & CPUE/LPUE
  - Rates of Biological Sampling of the Catch
  - Discard Estimates
- **Survey Calibration Studies**
  - Vessel/Door/Net Changes over Time
  - Estimation of Calibration Coefficients
  - Effects of Alternative Estimation Methods

Generic Assessment Issues II

Uncertainty in Assessments
- < Bootstrap Estimates of Precision in ADAPT
- < Sensitivity of Assessment Results to Potential Biases
- < Explicit vs. Implicit Sources of Uncertainty
- < Conveying Uncertainty to Managers

Medium-Term Projections
- < Utility of Results
- < Stock Rebuilding Targets
- < Estimation Methods
- < Presentation of Results

Georges Bank Haddock
Research Recommendations

- *Improve biological sampling of U.S. commercial landings and discards.*
- *Examine the effects of large tows on overall and age-specific abundance indices, specifically with reference to closed areas.*
- *Examine effects of abrupt changes in mean weights during the 1990s, specifically with respect to the 1989-1991 year classes in the eastern part of Georges Bank.*
- *Investigate factors associated with apparent recent improvements in survival ratios (R/SSB).*

Georges Bank Cod
Issues in Assessment

- Data Used:
  - ANALYTICAL, Age structured assessment
  - ADAPT VPA Calibration of age 1-10+catch at age, 1978-1996
  - NEFSC spring and autumn RV indices (1978-1996)
  - Canadian DFO spring RV indices (1986-1997)
  - CAA estimated for recreational landings 1981-1996
  - Discard ratios estimated for 1989-1996
- CAA:
  - *Commercial*: sufficient age/length data not available prior to 1978
  - 1994-1996 sampling poorest since 1982, notably in "large" market category.
  - 1996 "scrod" and "market" samples greater than 1995 samples
  - *Recreational*: derived from very few length samples, 1981-1996
- Sensitivity Runs:
  - Recreational (1981-1996): slightly higher stock size, with similar and SSB as base run
  - LPUE (1978-1996): lower stock size and higher F than final run
- Other issues:
  - Transboundary stock: use Canadian CAA and RV survey
  - Canadian RV survey indices for 1993 and 1994 not used (lack of coverage)
  - LPUE not used as in previous assessments (1994-1996 data unreliable)
  - Discards not incorporated: lack of samples, 1978-1988
  - Recreational not used: poor sampling, lack of samples, 1978-198

Georges Bank Cod
Assessment Summary

< Current landings (1996) = 8,900; up from 7,900 mt in 1995
< Survey indices fluctuate near record-low values
< LPUE declined by 1993 to the lowest LPUE in time series
< VPA results
  - Current F(1996) = 0.17
  - Current SSB(1997) = 46,400 t
  - Shift to older ages in PR
  - Mean F (4-8,u) declined from record-high in 1994 to record-low in 1996
  - Record-low recruitment of the 1994-1996 year classes
< Yield and SSB per Recruit Model
  - $F_{0.1} = 0.17$
  - $F_{max} = 0.32$
  - $F_{20\%} = 0.43$
< Projections
  - Short term:
    At status quo F, landings in 1998 (8,400 t) remain stable, SSB increases 35% during 1996 (41,200 t) to 1999 (55,000 t)
    Higher F scenarios initally increase landings and SSB, with subsequent declines
  - Medium term: 1998 to 2006
    Landings increase from 8,200 mt to 29,400 mt
    SSB improves from 53,700 mt to 199,900 mt
    Median recruitment increases from 14 million to 34.4 million
< CONCLUSIONS
  - Stock remains in overexploited state with low biomass
  - Recruitment of three most recent year classes at record low
  - SSB remains near record low value
  - Current F(96) nearly equal to F0.1

Georges Bank Cod Research Recommendations

< Further investigate the effect of closed areas on the use of LPUE as an index of abundance, specifically examine the impact of change in fleet distribution as a result of progressive exclusion from the Canadian zone and then from Closed Area II.

< Further investigate the basis for deriving the recreational component of the cod catch. Specifically, the effect of sampling levels in the party and charter categories at age can be used to augment the commercial landings at age in the VPA.

< Further examine discard rates in years prior to 1989 before discard data can be incorporated into the catch at age.

Gulf of Maine Cod Assessment Issues — I Data Used

< CURRENT ASSESSMENT
  - ANALYTICAL (Age Structured)
  - ADAPT VPA Calibration of Age 2-7+ CAA, 1982-1996
  - ANALYTICAL (Age Structured)
  - ADAPT VPA Calibration of Age 2-7+ CAA, 1982-1996
< 1996 landings were 7,200 t; up from 6,800 t in 1995
< Sample summary
  - 1994-1995 sampling improved over 1993, but low re: landings
  - 1996 much improved
< Catch at Age
  - Reduction in proportion of older 9>7 yr) ages and few fish >9 yr after 1990
  - 1992 year class 62% by number, 57% by weight in 1996
< Commercial LPUE
  - LPUE declined in 1993 to pre-1989 level; not used in 1994-1996
< Survey Indices
  - Low number and weight per tow persist through autumn 1996
  - Low recruitment indices in NEFSC and Massachusetts DMF surveys in 1995 and 1996
  - Survey Zx 1.1 in 1991-1993 and 1994-1996
< VPA Calibration
  - Unusual F pattern in 1994 for fully recruited ages 4 and 5
  - Record low recruitment of 1994 and 1995 year classes <1 million fish

Gulf of Maine Cod Assessment Issues — II
Quality of CAA

Commercial CAA
- < Catch at Age
  - Majority of catch at ages 3 and 4 (by number) and ages 4 and 5 (by weight)
  - Reduction in proportion of older (>7 yr) ages and few fish > yr after 1990
  - 1992 year class 62% by number, 57% by weight in 1996
  - High mean wts. at age in 1995 at ages 5 and 6 ==> Low Numbers may have influenced high F at age 4 and 5 in 1994
- < Sampling Intensity
  - Poor sampling in 1993
  - 1994-1995 sampling improved, but low for older ages (Large market cateogory)
  - Overall improvement in 1996

Recreational CAA
- < Catch at Age
  - Length composition indicates higher proportion of smaller fish compared to commercial catch
  - Majority of catch at ages 2 and 3 (by number) and ages 3 and 4 (by weight)
  - Note: Commercial age/length key used to distribute numbers at length to numbers at age
- < Sampling Intensity
  - Generally fewer than 1,000 (often <500) fish measured per year
  - Length data (and age) samples pooled on annual basis

Gulf of Maine Cod Assessment Issues — III
Alternative Runs

- < Six alternative assessment models using Stock Synthesis Approach (Fournier and Archibald, 1982)
  - Employed same survey indices as ADAPT
  - Included commercial CAA alone and commercial and recreational CAA
  - Lognormal, multinomial, and robust error structure assumptions
- < Lognormal model results
  - Log-normal error structure most closely approximated ADAPT/VPA
  - 1996 and 1997 Fs for fully recruited ages were estimated to be approximately 1.0
  - Recent recruitment (1994 and 1995 year classes) poorest ever in both analyses
  - SSB declining sharply since 1989, remaining lowest ever through 1997
- < Multinomial model results
  - Better fit to older ages and slightly lower Fs
  - Lower Fs at older ages in earliest years
  - Slightly lower Fs in recent years at all ages
  - Slightly better recruitment of 1994 and 1995 year classes
  - Marked increase in catchability by the fishery over time
  - Robust likelihood results similar to multinomial, but with very high Fs at older ages since 1988
- < Conclusions
  - All model formulations produce results similar to ADAPT/VPA:
    1) Sharp decrease in SSB since 1989
    2) Poorest recruitment on record for 1994 and 1995 year classes
    3) F on fully recruited ages has remained at or above 1.0 since early 1980s

Gulf of Maine Cod Assessment Issues — IV Sensitivity Runs

- Four Sensitivity Runs (including final)
  - Two with commercial CAA alone
  - Two including recreational CAA
- VPA results
  - 1996 terminal F for fully recruited ages ranged from 0.96 to 1.04 (essentially 1.0)
  - Recruitment estimates slightly higher when recreational CAA included
  - Recent recruitment (1994 and 1995 year classes) poorest ever in all runs
  - Retrospective patterns in recruitment not persistent
  - SSB remains low in all runs
- Final VPA calibration used commercial CAA alone due to
  - poor sampling of recreational CAA
  - uncertain allocation of recreational cod catch between Gulf of Maine and Georges Bank

Gulf of Maine Cod Assessment Summary

- Current landings (1996) = 7,200 t
- LPUE declined sharply in 1992, remained low in 1993-1996
- Survey Indices remain close to record-low levels in 1995-1996
- VPA results
  - current F(1996) = 1.04
  - current SSB(1996) = 9,200 t
- Yield and SSB per Recruit Model
  - $F_{0.1} = 0.16$
  - $F_{max} = 0.29$
  - $F_{20\%} = 0.37$
- Projections
  - At status quo F, SSB declines to unprecedented record low
  - Landings decline in 1998 under all F scenarios
  - SSB increases over the medium term at Fmax (0.29) to 13,000 t
- CONCLUSIONS
  - The stock is overexploited, current F(1996) is well above Fmax and about 3X F20%
  - Spawning stock biomass is at a record low level and is projected to decline further in 1998 and 1999

Gulf of Maine Cod Research Recommendations

- Further investigation of the changes in effort and LPUE in the VTR data set is required before LPUE can be used to calibrate VPA.
- Further investigation of the basis for deriving the recreational component of the cod catch, specifically the effect of sampling levels in the party and charter categories, is required before the recreational landings at age can be used to augment the commercial landings at age in the VPA.
- Further examination of discard rates in years prior to 1989 is required before discard data can be incorporated into the catch at age.

SNE Yellowtail Flounder Assessment Issues

- <u>Commercial Landings 1973-1996</u>
  - Collect data for quarterly estimation
  - Per capita sampling adequate 1973-1996
- <u>Commercial Discards 1973-1996</u>
  - Need adequate sea samplin
  - VTR discard ratios
- <u>Survey Indices 1973-1996</u>: Spring, Autumn, Scallop

SNE Yellowtail Flounder Assessment Summary

- **Fishing Mortality:** $F_{0.1} = 0.27$, $F_{96} = 0.12$, below reference point in 1996
- **Spawning Stock Biomass:** SSB-threshold = 10,000 mt, $SSB_{97} = 5,000$ mt, below threshold, but increasing
- **Short-term:** stock improving, could reach 8,000 mt SSB by 1999 at $F = 0.12$
- **Medium-term:** stock should rebuild if fishing rate remains low

SNE Yellowtail Flounder Research Recommendations

- Improve sea sampling coverage on trawl and scallop vessels.
- Increase sampling frequency on research surveys.
- Improve commercial length and age samples.
- Examine VTR data for mesh-specific discard ratios.
- Evaluate changes in maturity at age in recent years.
- Evaluate performance of the scallop survey as a tuning index.
- Evaluate winter survey as a tuning index.

Georges Bank Yellowtail Flounder Issues in the Assessment

- **Data in the assessment:** 1973-1996 U.S. and Canadian commercial landings and discards (recreational catch is negligible), U.S. spring and fall trawl survey indices, U.S. scallop survey indices, and Canadian trawl survey indices.
- **Quality of Catch at Age:**
  - Age composition by quarter and statistical area not available 1994-1996. Catches characterized semiannually for the entire stock area.
  - Sexually dimorphic growth not considered in estimates of catch at age.
- **Sensitivity Runs:** Log-transformed survey indices and removal of uncertain age-1 discard estimates. Results and diagnostics similar among all sensitivity runs.
- **Alternative Methods:** Biomass dynamics model (ASPIC) using total catch (1963-1996), U.S. spring and fall survey indices, and the Canadian survey index of biomass. Magnitude and temporal pattern of mean biomass and F similar to those from VPA.

Georges Bank Yellowtail Flounder Assessment Summary

- **F and SSB Trends:**
  - F averaged 1.2 during 1973-1994 and decreased to 0.1 in 1996
  - SSB was 21,000 mt in 1973, declined to less than 4,000 mt during 1984-1988, fluctuated below 6,000 mt from 1989 to 1994, and increased to 11,700 mt in 1996
- **Probability Distributions for F and SSB in 1996:**
  - 80% chance that F was between 0.08 and 0.14, and nearly 0% probability that F exceeded $F_{0.1}$ (0.25)
  - 80% chance that SSB was between 9,800 and 14,600 mt, and 12% probability that SSB was below the rebuilding threshold of 10,000 mt
- **Projections:**
  - At $F_{0.1}$, landings and SSB will continue to increase in the next three years. At F_{96}, landings will decrease slightly in 1997, then increase in 1998 and 1999, and SSB will continue to increase in the next three years.
  - At $F_{0.1}$, landings will increase to 8,400 mt in 2006, and SSB will increase to 46,200 mt in 2006. At F_{96}, landings will increase to 5,500 mt in 2006, and SSB will increase to 71,600 mt in 2006.

Georges Bank Yellowtail Flounder Research Recommendations

- *Extending the VPA back to the 1960s should be explored.*
- *Changes in maturity should be closely monitored.*
- *The NEFSC winter survey should be modified to ensure coverage of Georges Bank.*
- *Evaluate the feasibility of sex identification in all field sampling to estimate catch at age and survey indices by sex.*
- *The number of ages in the VPA should be expanded.*

	STOCK LEVEL		
EXPLOITATION STATUS	LOW	MEDIUM	HIGH
OVER EXPLOITED	REDUCE EXPLOITATION, REBUILD STOCK	REDUCE EXPLOITATION, BROADEN AGE DISTRIBUTION	REDUCE EXPLOITATION, INCREASE YIELD PER RECRUIT
FULLY EXPLOITED	REDUCE EXPLOITATION, REBUILD STOCK LEVEL	MAINTAIN EXPLOITATION RATE AND YIELD	MAINTAIN EXPLOITATION RATE AND YIELD
UNDER EXPLOITED	MAINTAIN LOW EXPLOITATION WHILE STOCK REBUILDS	INCREASE EXPLOITATION SLOWLY	INCREASE EXPLOITATION, REDUCE STOCK LEVEL

Stock Status Summary

	Current Fishing Mortality	1994-1996 Recruitment	1994-1996 SSB	Biomass Threshold
Gulf of Maine Cod	Well Above Target	Low	Low/ Declining	Threshold Not Defined
Georges Bank Cod	Near Target	Low	Low/ Increasing	Well Below
Georges Bank Haddock	Below Target	Low	Low/ Increasing	Well Below
Georges Bank Yellowtail	Below Target	Average	Low/ Increasing	Near
SNE Yellowtail	Below Target	Low	Low/ Increasing	Below

SARC 24 ADVISORY OVERVIEW

< The Situation for Georges Bay Cod, Haddock & Yellowtail, and SNE Yellowtail Has Improved; Gulf of Maine Cod is Unchanged

< Except for GM Cod, F Has Been Reduced BELOW Overfishing Reference Points & is NEAR Rebuilding Fs

< Except for GM Cod, Some Rebuilding of SSB has Occurred. GB Yellowtail is near SSB Threshold, but SSB for all Stocks is LOW relative to Historical Levels & BMSY

< Recruitment in Recent Years Remains LOW Relative to Historic Levels

< Short-Term Projections Indicate Maintenance or Modest Increases in SSB at Current Fs, Except for GM Cod

< Strong Management Measures Are Warranted for GM Cod to Reduce the Risk of Stock Collapse

< Continued Rebuilding of Four Stocks Will Be Jeopardized if Fishing Mortality Rates Are Allowed to Increase

< Efforts to Reduce F for GM Cod Should NOT Come at the Expense of Other Heavily Exploited Stocks in the Gulf of Maine or Elsewhere

Probability SSB ≥ Threshold by 2004
Based on Two Sets of Projections

	Threshold (kmt)	Amendment 7	SARC 24
Georges Bank Cod	70	0.76	1.00
Georges Bank Haddock	80	0.27	0.43
Georges Bank Yellowtail	10	0.98	1.00
SNE Yellowtail	10	0.94	1.00

Appendix E

Glossary

ABC (acceptable biological catch): Maximum amount of fish stock that could be harvested without adversely affecting recruitment or other biological components of the stock. The ABC level is typically higher than the total allowable catch.

ACFM (Advisory Council on Fisheries Management): An ICES advisory body. ACFM serves as a review panel, formulating advice based on assessments and draft advice from area-based assessment working groups. In addition to the assessment working groups, ACFM coordinates the work of methodologically oriented working groups.

ADAPT (ADAPTive framework): Age-structured assessment model used to estimate the abundance of fish stocks. The ADAPT technique is used to tune or modify the VPA model to minimize the discrepancy between empirical and simulated values of variables.

Amendment 4: Amendment to the Northeast Multispecies Fisheries Management Plan enacted in 1991 that extended restrictions on minimum mesh sizes and closed areas for a number of groundfish species.

Amendment 5: Amendment to the Northeast Multispecies Fisheries Management Plan enacted in 1994 that required extensive new regulations on minimum mesh sizes, minimum size limits, closed areas, new fishing permits, and logbook reporting. This amendment had the objective of reducing the fishing effort on cod, haddock, and yellowtail flounder stocks by 50% over five years (1999) based on 1993 levels. Amendment 5 established a days at sea program to reduce fishing effort.

Amendment 6: Amendment to the Northeast Multispecies Fisheries Management Plan enacted in 1994 as a secretarial amendment to the plan to extend emergency regulations concerning the possession limit for haddock, limits on scallop vessels landing haddock, and modifications to closed-area regulations.

Amendment 7: Amendment to the Northeast Multispecies Fisheries Management Plan enacted in 1996 to broaden and reinforce regulations enacted under Amendment 5. Amendment 7 extended closed areas and times, minimum size limits, set target TACs for species, established fishing mortality goals, set up a multispecies monitoring committee, and increased the fishing effort reduction goals to a 50% reduction within two years (1997) based on 1993 levels.

BMV: Type of trawl door used in early surveys by the United States. The BMV design, originally of Norwegian origin, was replaced by Portuguese polyvalent doors in 1985.

CAFSAC (Canadian Atlantic Fisheries Scientific Advisory Committee): Advisory body formed in 1977 to provide peer-reviewed scientific advice for the management of Canadian groundfish stocks.

Catchability/availability: Vulnerability of fish to capture by survey gear or fishing gear. The behavior, location, size, and abundance of fish at certain times may affect their vulnerability to capture. The type of vessel or fishing gear used and the competence of a fishing crew may also affect catchability.

Catch at age: Number or mass of fish from each cohort captured by a fishery for each year.

Coefficient of variation: Degree of variation, or distribution, of probabilities around a mean value. A high coefficient of variance is indicative of wide variation in the data being analyzed. Coefficient of variation is typically measured as percentage deviation above and below a mean value.

Cohort: Fish born in a given year. (see Year class).

Compensation: Decrease in number of recruits per spawner as spawning stock biomass increases.

CPUE (Catch per unit effort): Weight of fish harvested for each unit of effort expended by vessels in the fishery. CPUE can be expressed as weight of fish captured per fishing trip, per hour spent at sea, or through other standardized measures.

Days absent: The amount of time, or number of days, a vessel is away from port. Time spent fishing, steaming to fishing grounds, or other non-fishing activities are included.

DAS (days at sea): Fishery effort control program that establishes a limit on the number of fishing days that a vessel may participate in fisheries included in the Northeast Multispecies Fishery Management Plan. The number of days a vessel may fish is determined based on past participation in the fishery (individual days at sea), or a predetermined value for the fleet (fleet days at sea). The amount or time that qualifies as a day at sea includes time spent engaged in fishing or fishing related activities.

Depensation: A reduction in per capita productivity at low stock sizes.

Depensatory model: A model that exhibits depensation in the spawner-recruit relationship.

DFO (Department of Fisheries and Oceans): Federal agency in Canada responsible for management of fisheries in Canadian federal waters.

EEZ (exclusive economic zone): Zone extending from the shoreline out to 200 nautical miles in which the country owning that shoreline has the exclusive right to conduct certain activities such as fishing. In the United States, the EEZ is split into state waters (typically from the shore out to 3 nautical miles) and federal waters (typically from 3 to 200 nautical miles).

EFI (exploitation fraction index): The ratio of the catch to the survey index of abundance. This index is independent of the assessment model, being based only on data. It is an index because the survey index is used rather than a survey estimate of total abundance.

F (fishing mortality): Instantaneous mortality rate due to fishing. F combined with the natural mortality rate (M) is the total instantaneous mortality rate for a given stock (Z).

$F_{0.1}$: Fishing mortality rate at which an increase in fishing mortality produces a 10 percent increase in yield per recruit relative to the first unit of effort on the unexploited stock (i.e., the slope of the yield-per-recruit curve for the $F_{0.1}$ rate is only one-tenth the slope of the curve at its origin).

F_{max}: Rate of fishing mortality that produces the maximum yield per recruit; The point beyond which growth overfishing occurs.

F_{MSY}: Fishing mortality rate that produces the maximum sustainable yield.

Fishery Management Councils: Eight regional fishery management councils are mandated in the Magnuson-Stevens Fishery Conservation and Management Act responsible for developing fishery management plans for fisheries in federal waters. Councils are composed of voting members from NMFS, state fishery managers, and individuals selected by governors of the coastal states. Nonvoting members include the U.S. Coast Guard, the U.S. Fish and Wildlife Service, and other federal officials. Regional councils exist for the Caribbean, Gulf of Mexico, Mid-Atlantic, New England, North Pacific, Pacific, South Atlantic, and the Western Pacific regions.

FMP (fishery management plan): Management plan for fisheries operating in the federal EEZ produced by regional fishery management councils and submitted to the Secretary of Commerce for approval. These plans must meet certain mandatory requirements in the Magnuson-Stevens Fishery Conservation and Management Act before they can be approved or implemented.

FRCC (Fisheries Resource Conservation Council): Committee composed principally of fishing industry representatives and academics nominated by the Canadian Minister of Fisheries and Oceans. The FRCC provides management advice to the Minister.

Growth overfishing: Level of fishing mortality at which fish are captured before they achieve the maximum yield in mass due to growth. Growth overfishing occurs when fishing mortality is greater than F_{max}.

ICES (International Council for the Exploration of the Sea): International body established in 1902. ICES is a scientific forum for the exchange of information and ideas on the sea and its living resources and for the promotion and coordination of marine research by scientists in its member countries. Membership has increased from the original 7 countries in 1902 to the present 19 countries.

ICNAF (International Commission for Northwest Atlantic Fisheries): Fishery management organization founded by the United States and Canada in 1949 for joint scientific and management measures affecting certain groundfish stocks. ICNAF later evolved into NAFO.

Interim measures: Effort control measures enacted by the NEFMC in the early and mid 1980s to replace the FMP enacted in 1978. These measures included indirect controls of fishing effort such as minimum mesh and fish size restrictions.

ITQ (Individual Transferable Quota): Fishery management tool used in eastern Canada and other parts of the world that allocates a certain portion of the TAC to individual vessels, or fishermen, based on initial qualifying criteria. This allocation can be transferred or sold.

IWC (International Whaling Commission): International commission responsible for regulation of commercial, subsistence, and scientific whaling among member countries. IWC also conducts stock assessments of whale stocks and establishes quotas for subsistence and scientific harvests.

LPUE (landings per unit effort): Means of quantifying the CPUE. LPUE is the amount, or biomass, of fish landed per given unit of measure, typically measured on a per-trip or per-day basis.

M (natural mortality): The instantaneous mortality in a fish stock caused by predation, pollution, or other factors not related to fishing.

MMC (Multispecies Monitoring Committee): Committee established under Amendment 7 to review the amendment's effectiveness in meeting fishing mortality and stock rebuilding goals. The MMC is composed of staff from fishery management councils, NMFS scientists, state fishery officials, and industry representatives.

MRFSS (Marine Recreational Fisheries Statistics Survey): Primary source of marine recreational data for the New England region. MRFSS is operated by NMFS with the cooperation of coastal states. MRFSS is a design-based survey that produces estimates of total effort and catch in directed recreational fisheries.

M-SFCMA (Magnuson-Stevens Fishery Conservation and Management Act): Federal legislation responsible for establishing the regional fishery management councils and the mandatory and discretionary guidelines for federal fishery management plans. This legislation was originally enacted in 1976 as the Fishery Management and Conservation Act; its name was later changed to the Magnuson Fishery Conservation and Management Act, and in 1996 was renamed the Magnuson-Stevens Fishery Conservation and Management Act.

MSY (maximum sustainable yield): Largest average catch that can be captured from a stock under existing environmental conditions on a sustainable basis.

NAFO (Northwest Atlantic Fisheries Organization): Multinational fishery management organization. Since 1979, NAFO has regulated the harvesting of certain groundfish stocks in the NAFO Regulatory Area outside Canada's EEZ. NAFO establishes total allowable catches and sets quotas and conservation measures. The 15 member states of NAFO are Bulgaria, Canada, Cuba, Denmark (for the Faroe Islands and Greenland), Estonia, the European Union (including Spain and Portugal) Iceland, Japan, Korea, Latvia, Lithuania, Norway, Poland, Romania, Russia, and the United States.

NEFMC (New England Fishery Management Council): One of eight regional councils mandated in the Magnuson-Stevens Fishery Conservation and Management Act to develop management plans for fisheries in federal waters. NEFMC develops fishery management plans for fisheries off Maine, New Hampshire, Massachusetts, Connecticut, and Rhode Island. It is comprised of voting members from NMFS, state fishery managers, and individuals selected by governors of the five states. Nonvoting members include the U.S. Coast Guard, the US Fish and Wildlife Service, and other federal officials.

NEFSC (Northeast Fisheries Science Center): Regional science center operated by NMFS to conduct fishery science in the northeastern United States, including the NEFMC region.

NMFS (National Marine Fisheries Service): Federal agency within NOAA charged with overseeing the management and regulation of federal fisheries.

NOAA (National Oceanic and Atmospheric Administration): Federal agency within the Department of Commerce responsible for overseeing oceanographic and atmospheric science and regulation. NMFS is part of NOAA.

R (recruitment): Number, or percentage, of fish that survive from birth to a specific age or size. The specific size or age at which recruitment is measured may correspond to when the fish first become vulnerable to capture in a fishery or when the number of fish in a cohort can be estimated reliably by stock assessment techniques.

R/SSB (recruitment/spawning stock biomass): Number of fish recruited into a fishery from a given mass of spawning fish, usually expressed as number of recruits per kilogram of mature fish in a given stock. This ratio can be computed for each year class and is often used as an index of prerecruit survival. High R/SSB in a given year indicates an above-average recruitment of young fish per unit of biomass compared to other year classes.

RAP (Regional Advisory Process): Canadian stock assessment review process involving federal, provincial, and academic scientists, and possibly fishery managers, fishers, and processors. RAP generates a consensus stock status report presented to the FRCC.

Recruitment overfishing: Level of fishing mortality at which the recruitment of fish to the spawning stock biomass is significantly reduced. Recruitment overfishing is characterized by a decreasing proportion of older fish in the fishery and consistently low average recruitment over time.

SARC (Stock Assessment Review Committee): U.S. committee composed of scientists from NMFS, the appropriate regional council, and other academic scientists. SARC reviews stock assessments and related documents produced in the SAW, develops management advice, and agrees on working papers to be published.

SAW (Stock Assessment Workshop): Workshop responsible for producing the initial stock assessment documents used for formulating management advice. SAW is composed of scientists from NMFS, DFO, and NEFSC involved in conducting the surveys and other data analysis used in stock assessment reports.

SNE: Southern New England.

Appendix E

SPA (sequential population analysis): Retrospective analysis of the number of fish alive in each cohort for each past year. SPA relies on a relationship that the number of fish alive in each cohort at the beginning of next year can be calculated from the number alive this year reduced by the fishing mortality and the natural mortality of that cohort for the present year. This method requires catch-at-age data. SPA is similar to VPA; however, it allows fishing mortality and natural mortality to vary between years and with the age of the fish.

SSB (spawning stock biomass): Total mass of all sexually mature fish in a stock.

SSB/R (spawning stock biomass per recruit): Average expected contribution to spawning stock biomass for each recruit during its lifetime. SSB/R is calculated over the life span of the year class of the recruit according to a particular schedule of fishing mortalities. The exploitation pattern, rate of growth, and natural mortality rate of the recruit are usually assumed to be constant.

SSC (statistical and scientific committee): Fishery management advisory body composed of federal, state, and academic scientists that provides scientific advice to a fishery management council.

TAC (total allowable catch): Total catch permitted to be caught from a stock in a given time period, typically a year. In the United States, this level is determined by fishery management councils in consultation with NMFS. Typically, this level is lower than the acceptable biological catch.

VPA (virtual population analysis): Retrospective analysis of the number of fish alive in each cohort for each past year. VPA relies on a relationship that the number of fish alive in each cohort at the beginning of next year can be calculated from the number alive this year reduced by the fishing mortality and the natural mortality of that cohort for the present year. This method requires catch-at-age data.

Year class: Fish of a given species spawned or hatched in a given year (see Cohort); a three-year-old fish caught in 1998 would be a member of the 1995 year class.

Yield per recruit: Expected yield in mass from an individual fish over its life span. Yield per recruit is sometimes used to calculate ABCs and TACs.

Z (total mortality): Instantaneous mortality in a fish stock from all causes, combining fishing mortality and natural mortality: $Z = F + M$.

Appendix F

Extending Data Series and Alternative Projection Results for Gulf of Maine Cod

The committee utilised the services of a consultant, Dr. Graeme Parkes of MRAG Americas Inc., to (1) attempt to replicate the National Marine Fisheries Service (NMFS) assessments (described in Chapter 2), (2) investigate the possibility of extending stock assessments backward in time, and (3) provide alternative evaluation of management strategies by using different recruitment scenarios.

Extending Data Series Back In Time

Task 2 was an investigation of whether the survey data before the period covered by the stock assessments could be used in conjunction with ADAPT stock assessment results to provide historical estimates of recruitment and spawning biomass. The reason it cannot be done is that there is no reliable age composition information from the harvest prior to the assessment periods now used.

The survey abundance indices (I_a, mean catch per tow by age a as input into ADAPT) and calibration coefficients (unbiased estimates of q_a from ADAPT bootstrap results) were obtained. Estimates of absolute numbers at age N_a for the spring (1968 to 1996) and autumn (1963 to 1996) survey series were made using the relationship $N_a = I_a/q_a$.

These estimates were used to calculate spawning stock biomass by using mean weights at age and maturity ogives from ADAPT. No mean weight-at-age data were available for years prior to the periods covered by ADAPT, so the arithmetic mean of the weights at age was used for these years. The maturity ogive for the earliest year in the ADAPT was used for all earlier years in the survey series.

This procedure provided estimates of recruitment (numbers at age), total numbers, and spawning stock biomass in all years of the survey time series. Figures F1.A-F1.E show plots of recruitment over time for each of the five stocks compared to ADAPT estimates (estimates of total numbers and spawning biomass have similar patterns and hence are not shown). Trajectories for the spring and autumn surveys are shown separately. For Gulf of Maine cod, the age 1 index was not used in ADAPT tuning, and therefore no estimate of q was provided. As a proxy, a rough estimate was calculated as the average of I_1/N_1 for each survey series.

Recruitment estimates from application of this method are much more variable than those from ADAPT, showing the desirability of using an age-structured assessment model rather than converting survey estimates. Recruitment estimates from the period before the ADAPT assessments are generally higher on average than those during the assessment period, suggesting that recruitment, abundance, and biomass were higher prior to the period covered by the stock assessments.

Thus, better methods should be sought to extend assessments back in time in order to give a better historical perspective of the stocks. One method would be to construct or infer age frequencies from any commercial catch samples collected prior to the current assessment period. Then, assessment using ADAPT would simply employ data from the extended period. For Georges Bank haddock, samples go back to 1931, which permit a long-term assessment. The second method would be to use different assessment methods such as Stock Synthesis or Autodifferentiation (AD) Model Builder (see NRC, 1998, for a description of these models), which can take advantage of survey age composition data to extend the period without knowing the age composition of the commercial catch.

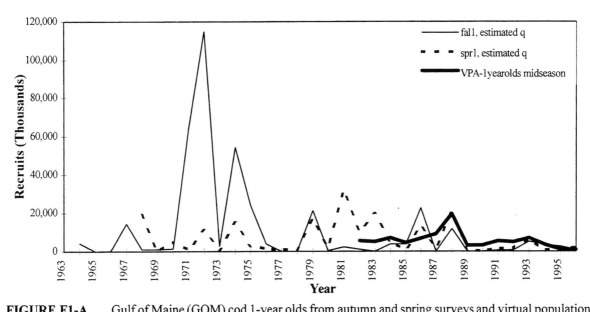

FIGURE F1-A Gulf of Maine (GOM) cod 1-year olds from autumn and spring surveys and virtual population analysis (midseason).

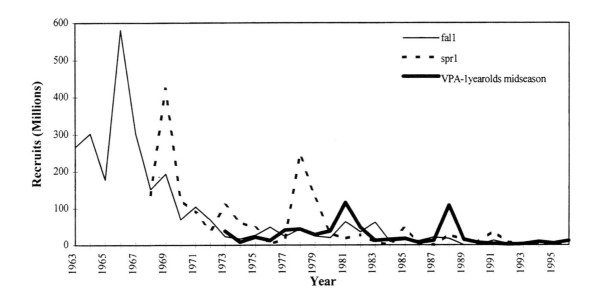

FIGURE F1-B Georges Bank (GB) cod 1-year olds from autumn and spring surveys and virtual population analysis (midseason).

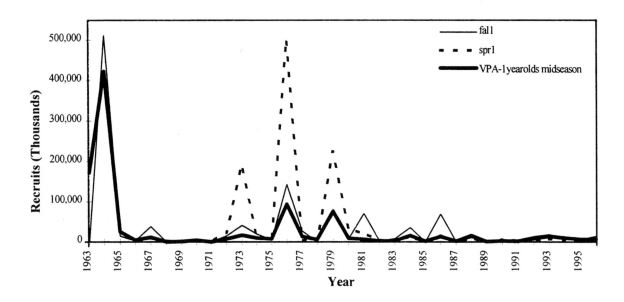

FIGURE F1-C Georges Bank (GB) haddock 1-year olds from autumn and spring surveys and virtual population analysis (midseason).

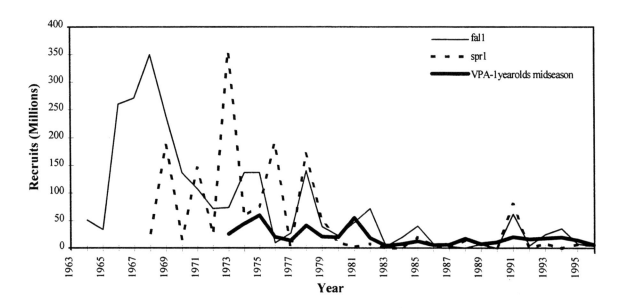

FIGURE F1-D Georges Bank (GB) yellowtail 1-year olds from autumn and spring surveys and virtual population analysis (midseason).

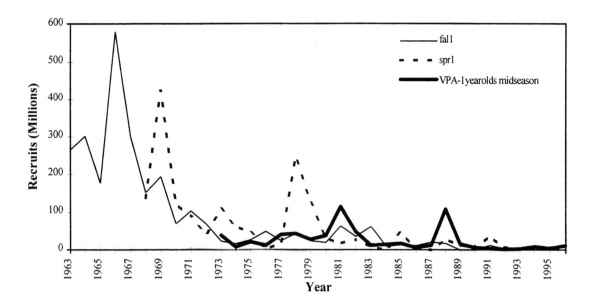

FIGURE F1-E Southern New England (SNE) yellowtail 1-year olds from autumn and spring surveys and virtual population analysis (midseason).

Alternative Projections

Task 3 was to develop alternative spawner-recruit relationships and use these to provide alternative forecasts of abundance over the standard 10-year period employed by NMFS. Initial projections were made by using a Beverton-Holt model, a constant recruitment model, and a density-independent model with recruitment proportional to spawning biomass. The alternative projections were stochastic so that probability distributions could be obtained. The first step in this analysis was to review the stock and recruitment data and Beverton-Holt parameter estimates presented in the SARC Working Paper Gen 2. The spawning stock-recruitment plots were regenerated for each stock from ADAPT results provided in the stock assessment papers. Spawning stock biomass (SSB) at the start of the spawning season in year t was matched with recruitment of 1-year olds at the start of year $t + 1$. A non-linear least-squares minimization routine (using Excel Solver, Systat, or Splus) was used to refit the parameters of the Beverton-Holt model to the ADAPT estimates in order to test the parameters estimated by NMFS. For Gulf of Maine cod, the estimated parameters and fitted curve were very different from those in Working Paper Gen 2, suggesting that it is difficult to get consistent parameter estimates with a Beverton-Holt model.

The stock-recruitment plots of data from ADAPT and fitted Beverton-Holt curves (NMFS and this study) are shown in Figures 2.1 through 2.5 of this report. These plots also show fit from the density-independent model.

A stochastic projection model was set up in Microsoft Excel, utilizing the add-in tool Crystal Ball, which enabled Monte Carlo simulations to be run and projection results presented in a spreadsheet format. This approach was favored to enhance the transparency of the method applied. After NMFS provided the ADAPT workspaces, the NMFS ADAPT bootstraps were rerun (see Chapter 2), which generated 200 vectors of stock abundance that were used as a seed for the projections. Rather than running 100 projections from each starting vector, as in the NMFS projections, a table lookup was generated which randomly selected one vector, with replacement, for each projection. A thousand

Other input data for the projections

SARC Working Paper Gen 2 indicated that selectivity at age (partial recruitment), weight at age, and maturity at age were "generally" taken as the averages of estimates for 1994-1996. The same approach was used for the review projections. For maturity at age, this normally meant taking the average of three identical age vectors. For weight at age, the averages over 1994 to 1996 of the values in the ADAPT input/output for catch weight at age at the beginning of the year were used.

For selectivity at age, there was a vector cited as an ADAPT input that featured a flat-topped selectivity curve, and there was also the matrix of estimated partial recruitments at age and year in the ADAPT output. In most cases, these seemed to show evidence of a domed partial recruitment. Literal interpretation of the NMFS description would have led to taking simple averages of these age vectors over the specified period of years, as was done for the catch weights at age. This did not seem right, however, because the ADAPT calculations were done under the effective assumption of flat-topped partial recruitment. Therefore the following procedure was used: (1) Assume that on input the partial recruitment was flat-topped from ages 4 to 9, with 10 being a plus group, with $F_{10} = F_9$ by assumption. This means that in the estimated partial recruitments in each year, average $(PR_4, \ldots +, PR_9) = 1.0$. (2) Average the partial recruitment estimates for each of ages 1, 2, and 3, and then assume a partial recruitment for each projection of $AvePR_1$, $AvePR_2$, $AvePR_3$, 1, 1, .. + . , .

After this procedure had been applied, it was noticed that the input vectors were cited for the yield-per-recruit calculations in one of the assessment papers. These values could have been used these if they were always cited, however, NMFS appeared to have followed a similar procedure to that outlined above, although the answers may not have been exactly the same. It would clearly have been very useful to have the input data included in the printout of the projection runs in SARC Working Paper Gen 2.

Output and Initial Projections

The spawning stock biomass was calculated at the start of the spawning season in accordance with the information provided in the ADAPT outputs (i.e., proportion of F and M before spawning). Projected recruitment values were constrained according to the methodology described in SARC Working Paper Gen 2 (using values of R/SSB_{10} and R/SSB_{90} for Gulf of Maine cod as amended in NEFSC (1997a, 27).

With five stocks, three recruitment models, and a variety of fishing mortality levels, the number of possible projections is very large. Projections were run with the Beverton-Holt model for each stock to determine whether the spreadsheet projection was providing results similar to the NMFS projection program (AGEPRO). The results were generally comparable with the projections presented in SARC Working Paper Gen 2. This paper evaluated the probability of rebuilding the spawning stock against agreed minimum threshold levels of spawning stock biomass. For other recruitment scenarios, the projections could be much different than those produced by NMFS, as illustrated in the following section.

Gulf of Maine Cod

After these initial projections, it was decided to conduct a limited set of simulations using Gulf of Maine cod, the only stock without an estimated decline in fishing mortality in the last two years. These projections were designed to illustrate how future stock responses to management can change for different stock-recruitment scenarios. The prior three recruitment scenarios were combined with an additional one reflecting the possibility of depensation, as defined below.

Appendix F 127

These models were postulated based on the assessment results:
1. the Beverton-Holt stock-recruitment model used in NMFS assessments;
2. recruitment will increase in proportion to stock size if spawning biomass is allowed to increase;
3. recruitment will stay constant on average at the historical mean value, independent of stock size; and
4. a depensatory spawner-recruit model in which the ratio R/S increases rather than decreases at small spawning biomass. As a result, a population at a very low biomass is likely to remain there for a long time until a positive recruitment event occurs. The spawner-recruit model is given by $R = S^g / (1/a + 1/k\, S^g)$. The parameter estimates of g, a, and k are 3.42, 8.6 x 10^{-11}, and 7857, respectively, obtained by least squares fit to recruit and spawner information from the ADAPT analysis.

In all cases, residual variance was assumed to be uncorrelated from year to year. Three levels of F were considered: 0.16 ($F_{0.1}$), 0.29 (F_{max}) and 1.04 (10-year mean). The four recruitment scenarios are illustrated in Figure F.2, along with the data points from the NMFS ADAPT results. Also indicated on this graph are levels of maximum R when SSB is greater than and less than the minimum observed level (in ADAPT results). When SSB is greater than the minimum observed level there is no practical limit on the level of R. When SSB falls below this level, however, R is restricted to values less than the lower of these two lines. In practice, this constraint had only a minor effect on projections except at the highest level of F (i.e., 1.04). The constraint on R was applied for all projections except those in which R is held constant, independent of stock size (scenario 3).

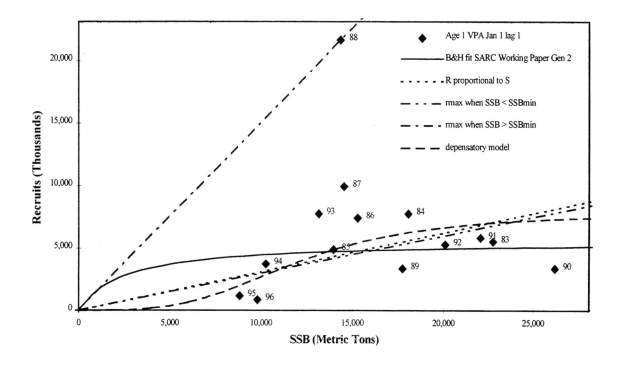

FIGURE F.2 Four recruitment scenarios for Gulf of Maine cod based on original NMFS VPA results 1983 to 1996.

The trend charts for spawning stock biomass, recruitment, and catch from the 12 projections (4 recruitment scenarios multiplied by 3 target Fs) are presented in Figures 2.9 through 2.12 of this report. An extra projection for R proportional to SSB at $F = 1.04$ was run without the constraint on R to demonstrate the effect that this constraint has at high levels of F (Figure 2.13). Discussion of the results of these projections is presented in the committee's report.

The results are summarised in a decision analysis table (Table 2.1 of this report), in which the consequences of alternative actions can be evaluated across different recruitment scenarios. To be even more useful in decisionmaking, such a table should be constructed using actual management tactics that could be employed to implement different target fishing mortalities rather than using the target fishing mortalities themselves.